Alan Bundy (Ed.)

Artificial Intelligence Techniques

A Comprehensive Catalogue

Fourth, Revised Edition

Springer

Editor

Alan Bundy
Department of Artificial Intelligence
The University of Edinburgh
80 South Bridge, Edinburgh EH1 1HN
Scotland, U.K.

E-mail: A.Bundy@ed.ac.uk

Assistant Editors

Alan Smaill
Ian Frank
Nicolas Nicolov

This is a completely revised edition of the "Catalogue of Artificial Intelligence Techniques", 3rd edition 1990, which appeared in the series "Symbolic Computation – Artificial Intelligence".

ISBN 3-540-59323-3 Springer-Verlag Berlin Heidelberg New York

ISBN 3-540-52959-4 3rd edition Springer-Verlag Berlin Heidelberg New York
ISBN 0-387-52959-4 3rd edition Springer-Verlag New York Berlin Heidelberg

Library of Congress Cataloging-in-Publication Data

Bundy, Alan.
Artificial intelligence techniques : a comprehensive catalogue / Alan Bundy (ed.). -- 4th rev. ed.
p. cm.
Rev. ed. of: Catalogue of artificial intelligence techniques. 3rd rev. ed. c1990.
Includes bibliographical references and index.
ISBN 3-540-59323-3 (softcover : alk. paper)
1. Artificial intelligence--Data processing. I. Bundy, Alan. Catalogue of artificial intelligence techniques. II. Title.
Q336.B86 1996 96-41821
006.3--dc20 CIP

Cover Design: design & production GmbH, Heidelberg
Typesetting: Camera ready by editor
SPIN: 10667278 45/3111 – 5 4 3 2 1 – Printed on acid-free paper

Preface

The purpose of "Artificial Intelligence Techniques: A Comprehensive Catalogue" is to promote interaction between members of the AI community. It does this by announcing the existence of AI techniques, and acting as a pointer into the literature. Thus the AI community has access to a common, extensional definition of the field, which promotes a common terminology, discourages the reinvention of wheels, and acts as a clearing house for ideas and algorithms. I am grateful to the impressive group of AI experts who have contributed the many descriptions of AI techniques which go to make up this Catalogue. They have managed to distill a very wide knowledge of AI into a very compact form.

The Catalogue is a reference work providing a quick guide to the AI techniques available for different tasks. Intentionally, it only provides a brief description of each technique, with no extended discussion of its historical origin or how it has been used in particular AI programs.

The original version of the Catalogue was hastily built in 1983 as part of the UK SERC-DoI, IKBS Architecture Study.[1] It was adopted by the UK Alvey Programme and, during the life of the programme, was both circulated to Alvey grant holders in hard copy form and maintained as an on-line document. A version designed for the international community was published as a paperback by Springer-Verlag. All these versions have undergone constant revision and refinement. Springer-Verlag has agreed to reprint the Catalogue at frequent intervals in order to keep it up to date. This is the fourth version of the Catalogue. Many experts in AI have collaborated to bring it up to date by adding many new entries and checking and revising all the old ones.

By 'AI techniques' we mean algorithms, data (knowledge) formalisms, architectures, and methodological techniques, which can be described in a precise, clean way. The Catalogue entries are intended to be non-technical and brief, but with a literature reference. The reference might not be the 'classic' one. It will often be to a textbook or survey article. The border between AI and non-AI techniques is fuzzy. Since the Catalogue is to promote interaction, some techniques are included because they are vital parts of many AI programs, even though they did not originate in AI.

[1] Mike Wilson, *Intelligent Knowledge Based Systems: A Programme for Action in the UK* (1983) (available from Mike Wilson, Informatics Division, Rutherford Appleton Laboratory, Didcot, Oxon. OX11 0QX).

We have not included in the Catalogue separate entries for each slight variation of a technique, nor have we included descriptions of AI programs tied to a particular application, nor of descriptions of work in progress. The Catalogue is not intended to be a dictionary of AI terminology, nor to include definitions of AI problems, nor to include descriptions of paradigm examples.

Entries are short (abstract length) descriptions of a technique. They include a title, list of aliases, contributor's name, paragraph of description, and references. The contributor's name is that of the original author of the entry. Only occasionally is the contributor of the entry also the inventor of the technique. The reference is a better guide to the identity of the inventor. Some entries have been subsequently modified by the referees and/or editorial team, and these modifications have not always been checked with the original contributor, who should not always be held morally responsible, and never legally responsible, for the final formulation.

The original version of the Catalogue was called "The Catalogue of Artificial Intelligence Tools" and also contained descriptions of portable software, e.g., expert system shells and knowledge representation systems. Unfortunately, we found it impossible to maintain a comprehensive coverage of either all or only the best such software. New systems were being introduced too frequently and it required a major editorial job to discover all of them, to evaluate them and to decide what to include. It would also have required a much more frequent reprinting of the Catalogue than either the publishers, editors or readers could afford. Also expert system shells threatened to swamp the other entries. We have, therefore, decided to omit software entries from future editions and rename the Catalogue to reflect this. The only exception to this is programming languages, for which we will provide generic entries.

New entries for the Catalogue or suggestions about its organisation or content are welcome, and should be sent to me (see page iv).

August 1996 Alan Bundy

Acknowledgements

I would like to thank:

- the SERC and DoI, who funded the initial production of this Catalogue as part of the Study of Architectures for Intelligent Knowledge-Based Systems;
- those members of the AI/IKBS community who wrote the individual entries;
- the review committee of the Catalogue, who vetted and improved them;
- my assistants who did the low level typing/editing work and tracked down many of the entries: Robin Boswell, Chris Dipple, Carole Douglas, Kave Eshghi, Ian Frank, Bob Fisher, Ian Gent, John Hallam, Mike Harris, Luis Jenkins, Helen Lowe, Seán Matthews, Santiago Negrete, Nicolas Nicolov, Mary-Angela Papalaskaris, Dave Plummer, Maarten van Someren, Andrew Stevens, Millie Tupman and Toby Walsh;
- John Taylor, for setting up the Architecture Study and for his unflagging enthusiasm and encouragement;
- Bill Sharpe, for being an ever present source of advice, help and information;
- John W.T. Smith for bearing the burden of the on-line Catalogue;
- Seán Matthews, who produced the typesetting code for the third and the current edition, using the TeX system with Leslie Lamport's LaTeX package and modified forms of Mike Spivak's AmsTeX and Donald Arseneau's citation code;
- the succession of assistant editors: Lincoln Wallen, Alan Smaill, Ian Frank and Nicolas Nicolov;
- and last, but not least, I would like to thank our publisher, Springer Verlag. Their considerable faith in both AI and the Catalogue is demonstrated by their committment throughout four editions. They have also been an invaluable source of detailed and helpful editorial comment and have shown amazing patience with us during the many production delays.

Alan Bundy

Guide to Using the Catalogue

There are two indices in this catalogue to help you find entries:
- A *Logical Table of Contents* which lists the techniques under various subfields of AI,
- An *Index of Definitions*; a topic/keyword index with main references underlined,

A *List of Contributors* is also included, giving the names of the people that have contributed descriptions of techniques.

The techniques are listed in alphabetical order, and numbered. The indices refer to these numbers rather than page numbers. The following conventions are used for the visual presentation:

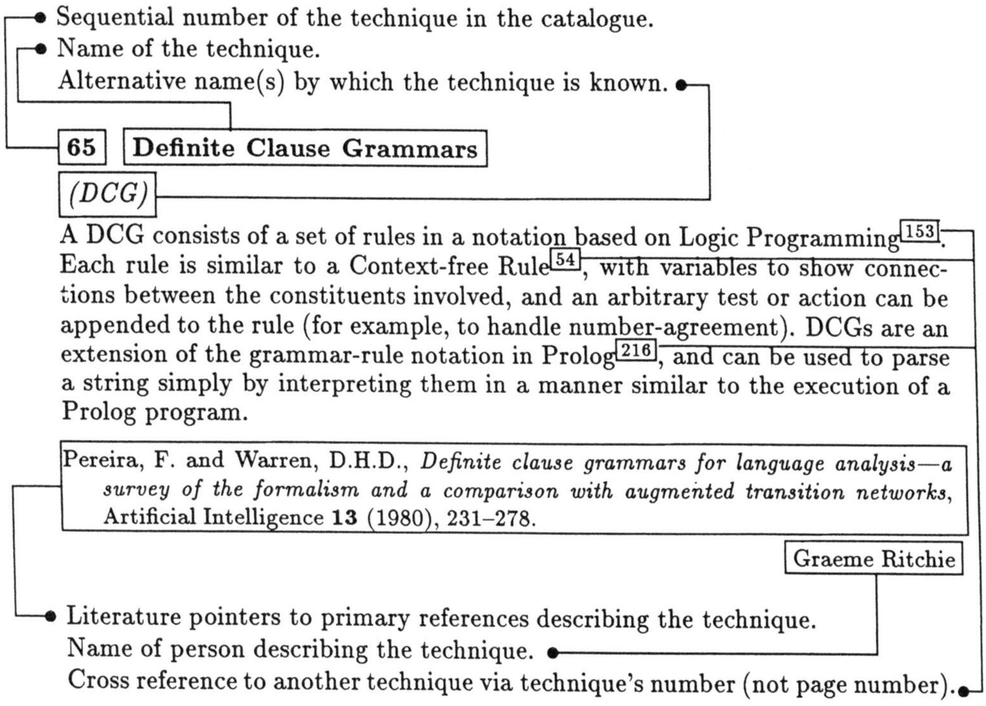

- Sequential number of the technique in the catalogue.
- Name of the technique.

Alternative name(s) by which the technique is known.

65 **Definite Clause Grammars**

(DCG)

A DCG consists of a set of rules in a notation based on Logic Programming [153]. Each rule is similar to a Context-free Rule [54], with variables to show connections between the constituents involved, and an arbitrary test or action can be appended to the rule (for example, to handle number-agreement). DCGs are an extension of the grammar-rule notation in Prolog [216], and can be used to parse a string simply by interpreting them in a manner similar to the execution of a Prolog program.

Pereira, F. and Warren, D.H.D., *Definite clause grammars for language analysis—a survey of the formalism and a comparison with augmented transition networks*, Artificial Intelligence **13** (1980), 231–278.

Graeme Ritchie

- Literature pointers to primary references describing the technique.

Name of person describing the technique.

Cross reference to another technique via technique's number (not page number).

Logical Table of Contents

Knowledge Representation

Learning

Logic Programming

Planning

Problem Solving

Programming Languages

Robotics

Search

Speech

Theorem Proving

Vision

1 $2\frac{1}{2}$-D Sketch

A viewer-centred representation making explicit the depths, local orientations and discontinuities of visible surfaces, created and maintained from a number of cues, e.g., Stereopsis[261] and Optical Flow[190]. It was thought by Marr to be at the limit of pure perception, i.e., subsequent processes are no longer completely data-driven, and for him it provides a representation of objective physical reality that precedes the decomposition of the scene into objects.

Marr, D., *Vision: A Computational Investigation into the Human Representation and Processing of Visual Information*, W.H. Freeman, San Francisco, CA, 1982.

T.P. Pridmore, S.R. Pollard, S.T. Stenton

2 A* Algorithm

A form of Heuristic Search[116] that tries to find the cheapest path from the initial state to the goal. Its characteristic feature is the evaluation function. This is the sum of two components: the estimated minimum cost of a path from the initial state to the current state, and the estimated cost from the current state to the goal. The first component can be calculated if the Search Space[260] is a tree, or it can be approximated by the cheapest known path if the search space is a graph. The second component must be defined, like any evaluation function, with respect to the domain. The heuristic power of this method depends on the properties of the evaluation function.

Hart, P.E., Nilsson, N.J. and Raphael, B., *A formal basis for the heuristic determination of minimum cost paths*, IEEE Transactions on SSC **4** (1968), 100–107 (a correction was published in SIGART Newsletter 37:28–29, 1972).

Korf, R.E., *Search: a survey of recent results*, Exploring Artificial Intelligence (Survey Talks from the National Conferences on Artificial Intelligence) (Shrobe, H.E., ed.), Morgan Kaufmann, San Mateo, CA, 1988, pp. 197–237 (Chapter 6).

Maarten van Someren

3 Abduction

Abduction was identified by the philosopher Charles Pierce as an especially important form of hypothesis formation. In the simplest case it has the form: from 'A' and 'B implies A' infer 'B'. The abductive hypothesis 'B' can be regarded as a possible explanation of 'A'. To be a useful hypothesis, 'B' should be consistent with other beliefs. Abduction is non-monotonic and non-deterministic. For example, from 'A', and 'B implies A', and 'C implies A' we can infer 'B' or we can infer 'C' as possible alternative hypotheses. Adding 'not B' as a new belief, non-monotonically withdraws 'B' as a hypothesis. In comparison to this logical view, some researchers like Reggia have taken Peirce's original idea and interpreted it in terms of set coverings. Abduction has recently gained popularity in Artificial Intelligence for such applications as fault diagnosis, language understanding and image understanding. It has also been proposed as an alternative to Non-monotonic Logics[182] for Default Reasoning[64].

Charniak, E. and McDermott, D., *Introduction to Artificial Intelligence*, Addison-Wesley, Reading, MA, 1985, pp. 21–22.

Peirce, C.S., Collected Papers of Charles Sanders Peirce, (Hartshorn, C., et al., eds.), vol. 2, Harvard University Press, Cambridge, MA, 1932.
Reggia, J A., Nau, D.S. and Wang, P.Y., *Diagnostic expert systems based on a set covering model*, International Journal of Man Machine Studies **19** (1983), 437–460.
Paul, G., *Approaches to abductive reasoning: an overview*, Artificial Intelligence Review **7** (1983), 109–152.

Robert Kowalski

4 Abstraction

Abstraction is a powerful heuristic used in many types of problem solving including planning, common-sense reasoning, theorem proving and learning. It usually involves a mapping from the original problem representation (often confusingly called the ground space) onto a new and simpler representation (often called the abstract space). One of the most famous uses of abstraction was in the ABSTRIPS planning system where the preconditions of Operators[187] were dropped according to their criticality, a numerical measure of their importance.

Plaisted, D.A., *Theorem proving with abstraction*, Artificial Intelligence **16** (1981), 47–108.
Sacerdoti, E.D., *Planning in a hierarchy of abstraction spaces*, Artificial Intelligence **5** (1974), 115–135.
Giunchiglia, F. and Walsh, T., *A theory of abstraction*, Artificial Intelligence **56** (1992), 323–390.

Toby Walsh

5 Active Vision

Active Vision is the philosophy that many robotics vision problems are greatly simplified if a robot is allowed to collect and analyse a sequence of images as it undergoes some appropriately chosen motion. Examples of such problems include Structure from Motion[266], Stereopsis[261], object segmentation and low-level reactive behaviours such as tracking and collision avoidance. Active vision systems are generally characterised by their reliance on fast real-time hardware, relatively simple image-processing algorithms, and little or no calibration. Position control is often by visual servoing: converting image movement directly into motor signals, and the systems usually retain only a small amount of internal state (for example, the mean and covariance representations of the Kalman filter). This branch of computer vision research may be seen as a reaction to approaches that concentrate on single, static images of scenes and insist on a full interpretation of the whole picture. Active systems generally perform impressively on tasks which have proved difficult for the more classical approach, but have not yet scaled to more complex problems such as scene understanding.

Aloimonos, J., Weiss, I. and Bandyopadhyay, A., *Active vision*, International Journal of Computer Vision **1** (1987), 333–356.
Special Issue on Active Vision, International Journal of Computer Vision **11** (1993), no. 2.

Andrew Fitzgibbon

6 Actors

The actor model of computation was developed by Hewitt in order to explore the fundamental issues involved with computation via message-passing. Actors are objects which know about other actors and can receive messages from other actors. Each actor is specified by detailing what kind of messages it will receive, and the sequence of actions of the actor, should it be sent one of these messages. Everything in an actor based programming system is an actor. To increment a value, for instance, a message would be sent to the actor representing the number, asking the actor to increment itself. The essential difference between the actor model of computation and the SmallTalk-80 language is that the sequencing of computations by actors depends critically on the concept of a 'continuation'. A continuation is an actor which is prepared to accept an (intermediate) value as a message, and continue the computation. In SmallTalk-80, an object (or actor) will instead return a message to the object which instigated the computation, in much the same way that a Pascal function will return a value to the routine which called it.

Hewitt, C., *Viewing control structures as patterns of passing messages*, Artificial Intelligence **8** (1977), 323–364, also in Artificial Intelligence: An MIT Perspective (Winston, P.H. and Brown, R.G., eds.), MIT Press, Cambridge, MA, 1979, pp. 433–465.

Mark Drummond

7 Agenda Based Systems

An agenda is a data structure containing tasks to be performed during a problem solving process, which may have attached justifications (for performing that particular task, see Truth Maintenance Systems[283]) and/or ratings of importance. Agenda based systems are a means of controlling inference, using some method of adding tasks to the agenda and of selecting, executing, and deleting tasks from the agenda. Tasks may be placed on the agenda simply as they are generated, either directly or as subtasks of existing tasks, with the agenda functioning simply as a stack. Otherwise, the usual control sequence for processing in agenda based systems, assuming tasks on the agenda are sorted by rating (importance), is:

1. Choose the most promising task from the agenda.
2. Execute this task.
3. If subtasks are generated then:
   - If a subtask is already on the agenda then add any new justifications (resulting from the current task) to its list of justifications.

     If a subtask is not on the agenda, then insert it, together with its justifications.
   - Compute the new ratings for all the tasks on the agenda.
4. Re-sort the agenda by rating.

Such systems essentially implement a Best-First Search[116] strategy with the rating of each node as its heuristic score. Examples of agenda based systems include DENDRAL, AM, CENTAUR, and KRL. Blackboard[23] systems also use agendas to control their operation.

Rich, E., *Artificial Intelligence*, McGraw-Hill, New York, 1991 (second edition).

Helen Lowe

8 Alpha/Beta Pruning

The Minimax[164] procedure need not be applied to every branch of the Search Tree[260]. Newell first observed that once a branch of a tree has been refuted, it need not be refuted again. The alpha-beta procedure was first formalised by McCarthy. Its properties are such that it is highly dependent on the order of presentation of alternatives. When alternatives are presented in the best order, it is possible to save $O(n^{\frac{1}{2}})$ of the total effort, where n is the number of leaf nodes in the tree. Recent advances in game searching, such as Iterative Deepening[138] and the use of Trans-Ref hash tables, have improved the performance of the alpha-beta search beyond the above specified bound. This is due to the fact that near perfect ordering can be achieved when the tree has many identical nodes in it, but in this case it can be more properly analysed as a graph.

Newell, A., Simon, H. and Shaw, J.C., *Chess playing programs and the problem of complexity*, Computers and Thought (Feigenbaum, E.A. and Feldman, J., eds.), McGraw-Hill, New York, 1963, pp. 39–70.

Korf, R.E., *Search: a survey of recent results*, Exploring Artificial Intelligence (Survey Talks from the National Conferences on Artificial Intelligence) (Shrobe, H.E., ed.), Morgan Kaufmann, San Mateo, CA, 1988, pp. 197–237 (Chapter 6).

Slate, D.J. and Atkin, L.R., *Chess 4.5—the Northwestern University chess program*, Chess Skill in Man and Machine (Frey, P.W., ed.), Springer-Verlag, Berlin, 1977, pp. 82–118.

Hans Berliner

9 Analogical Problem Solving

A technique whereby the current problem is solved by retrieving from memory a previously solved similar problem and appropriately adapting that problem's solution. Carbonell has developed a particular version of this technique based on Means/ends Analysis[159]. See also Case-based Reasoning[32].

Carbonell, J.G., *A computational model of analogical problem solving*, Proceedings of IJCAI-81, International Joint Conference on Artificial Intelligence, 1981, pp. 147–152.

Jim Doran

10 AND/OR Graphs

(*Game Trees*)

AND/OR graphs consist of nodes (representing goals and subgoals) connected by directed links—either AND links (representing subgoals that make up their parent goal) or OR links (representing alternative goals that satisfy the parent goal). AND/OR graphs are often used in theorem provers and game-playing programs. In game trees, the OR links from a node become the moves available to the player for whom a decision procedure is to be implemented (since it will be necessary to select exactly one of the alternatives if the game reaches that node). The AND links, on the other hand, represent the moves available to opponents, which cannot be controlled, therefore requiring a response to be considered for each branch. The leaf-nodes are often given a score which represents the payoff for one of the players if the game finishes in the corresponding state. If, as is often the case, the payoffs are zero-sum (i.e., one player's payoff is the negative of the other's) then these two players are called MAX (who tries to maximise his payoff by directing the play towards leaf nodes with high payoffs) and MIN (who tries to minimise MAX's return by directing the play towards leaf nodes with low payoffs), respectively. See also Minimax[164].

Nilsson, N.J., *Principles of Artificial Intelligence*, Tioga Pub. Co., Palo Alto, CA, 1980, and Springer-Verlag, Berlin, 1982.

Toby Walsh, Ian Frank

11 Associative Database

Associative database provides Pattern Directed Retrieval[196] by the 'Get Possibilities/Try Next' mechanism. The provision of context layers as in CONNIVER and QA4 allows for items to be associated with a value in some context.

Sussman, G.J. and McDermott, D.V., *From PLANNER to CONNIVER—a genetic approach*, Fall Joint Computer Conference, 1972, pp. 1171–1179.

Austin Tate

12 Associative Memories

Associative memories work by recalling information in response to an information cue, rather than an address (as in standard computer memory). Associative memories can be autoassociative—recalling the same information that is used as a cue, or heteroassociative—where the association is made with different information. The former can be used to complete a partial pattern, or clean up noisy data, the latter is more obviously useful as a memory. They can be implemented in many ways. For example Hopfield Nets[123] can be used for autoassociative memory, and Willshaw Nets[296] will do either.

Kohonen, T., *Self-organisation and Associative Memory*, (Springer Series in Information Sciences, vol. 8), Springer-Verlag, Berlin, 1989 (3rd edition).

Willshaw, D.J., Buneman, O.P. and Longuet-Higgins, H.C., *Non-holographic associative memory*, Neurocomputing: Foundations of Research (Anderson, J.A. and Rosenfeld, E., eds.), MIT Press, Cambridge, MA, 1988, pp. 151–156.

Martin Westhead

13 Augmented Transition Network

(ATN)

A representation for grammars developed from simple finite-state transition networks by allowing (i) recursion , and (ii) augmentation, i.e., the use of arbitrary tests and actions on arcs, giving full Turing machine power. The use of registers for storing constituents, and the use of tests and actions on register contents allow great flexibility in parsing, and in particular permit the construction of sentence representations quite distinct from the surface text, e.g., deep as opposed to surface syntactic structures. The form of grammar representation is procedurally oriented, but the grammar itself is separated from the interpretive parser, which is Top-down[279] and usually Depth-first[71]. ATNs can be adapted, for example, to guide parsing by explicit arc ordering. Problems arise with, for example, passing information between subnets, and the treatment of conjunctions.

Woods, W.A., *Transition network grammars for natural language analysis*, Communications of the ACM **13** (1970), 591–606, also in Readings in Natural Language Processing (Grosz, B.J., Sparck Jones, K. and Webber, B.L., eds.) Morgan Kaufmann, Los Altos, CA, 1986, pp. 71–87.

Karen Sparck Jones

14 Auto-epistemic Logic

Auto-epistemic logic is a logic for modelling the beliefs of agents who reflect on their own beliefs. It can express statements such as 'If I do not believe that p is true then q is true'. It characterises the beliefs that an ideally rational agent would be entitled to hold on the basis of such statements. The most distinctive feature of auto-epistemic logic is that it is a Non-monotonic Logic[182]: the set of basic beliefs can be modified with time. An agent who takes as a premise 'If I do not believe that p is true, then q is true' and who does not believe that p is true, will believe that q is true. If later he were to adopt the premise p as an additional belief then he would no longer be justified in believing that q holds. See also Logics of Knowledge and Belief[154].

Moore, R.C., *Auto-epistemic logic*, Non-standard Logics for Automated Reasoning (Smets, P., Mamdani, E.H., et al., eds.), Academic Press, London, 1988, pp. 105–136.

Colin Phillips

15 B* Algorithm

B* is a heuristic search method that can be applied to both adversary and non-adversary problems, but only when the search has an iterative character. It computes the best next step toward the solution on the basis of an evaluation function.

An evaluation function assigns two values to each node, a pessimistic and an optimistic value, (cf. Minimax[164]) on the basis of evaluation of the descendants. In non-adversary search this is done according to the following rules:

1. Evaluate the descendants of a node in arbitrary order. If one of the descendants of a node has a pessimistic value greater than the pessimistic value of its parent, then raise the pessimistic value of the parent to the value of this daughter. If a node has an optimistic value that is higher than that of any daughter, then lower it to the maximal optimistic value of its daughters.

2. Terminate when a daughter of the root of the search tree has a pessimistic value that is not lower than the optimistic value of all other daughters. The arc to that daughter is the best step to take.

In the case of adversary search (e.g., game playing), B* is the same as Alpha/Beta search[8], except that it stops once it has found the best next move. Best-first or Heuristic Search[116] may be implemented in this manner. B* is claimed to be a good model of the search of chess masters.

Berliner, H., *The B* tree search algorithm: a best-first proof procedure*, Artificial Intelligence **12** (1979), no. 1, 23–40, also in Readings in Artificial Intelligence (Webber, B.L., ed.) Tioga Pub. Co., Palo Alto, CA, 1981, pp. 79–88.

Maarten van Someren

16 Backwards Search

(*Goal-driven Search, Top-down Search, Backwards Chaining*)

A State Space[260] can be searched from the initial state to the goal state, applying Operators[187] forward (Forward Search[96]); alternatively the state space can be searched from the goal to the initial state applying the operators in reverse (backwards search).

Barr, A. and Feigenbaum, E.A. (eds.), *The Handbook of Artificial Intelligence*, vol. 1, William Kaufmann, Los Altos, CA, 1981.

Maarten van Someren

17 Bayesian Inference

(*Statistical Inference*)

Bayesian inference is one means by which knowledge based systems can reason when uncertainty is involved. Given a set of mutually exclusive hypotheses H_i and an evidence event E, we obtain from an expert estimates of the prior probabilities $P(H_i)$, and the conditional probabilities $P(E|H_i)$. Bayes' Rule then gives the probability of H_i given evidence E:

$$P(H_i|E) = c \cdot P(E|H_i) \cdot P(H),$$

where c is a normalising constant ensuring that $P(H_i|E)$ sum to unity. In practical problems E may be any subset of the set of all possible evidence events and the hypothesis of interest may be any subset of the set of all possible hypotheses. This tends to require a vast number of conditional probabilities to be calculated.

A solution is provided by techniques based on Bayesian Networks[18]. To specify such a network one need only estimate the conditional probabilities of each elementary event E_j given its immediate causes. When the network formed by these cause-effect relationships is loop free, the probability $P(H|E)$ for any subset E of evidence events and any subset H of hypotheses can be calculated by distributed message-passing techniques. When cycles are unavoidable, tree clustering and cycle cutset techniques (see Constraint Networks[51]) can be used to compute $P(H|E)$. An example of the use of Bayesian inference is the HUGIN system.

Andersen, S.K., Olesen, K.G., Jensen, F.V. and Jensen, F., *Hugin—a shell for building Bayesian belief universes for expert systems.*, Proceedings of IJCAI-89, pp. 1080–1085.

Pearl, J., *Probabilistic Reasoning in Intelligent Systems: Networks of Plausible Inference*, Morgan Kaufmann, San Mateo, CA, 1988.

Robert Corlett, Judea Pearl

18 Bayesian Networks

(*Causal Networks, Influence Diagrams*)

A Bayesian network is a graphical representation of probabilistic information, which uses directed acyclic graphs, where each node represents an uncertain variable and each link represents direct influence, usually of causal nature. This graphical representation is used to store probabilistic relationships, to control inferences, and to produce explanations.

To define a coherent probability distribution on all variables in the network, we must assess the conditional probabilities relating each variable to its parent variables. If the network formed by these cause-effect relationships is loop free, then inferences from evidence to hypotheses (e.g., finding the best explanation) can be performed in linear time, using message-passing techniques. When loops are unavoidable, tree clustering and cycle cutset methods (see Constraint Networks[51]) can be used to facilitate coherent inferences.

Data dependencies in Bayesian networks are identified by a graphical criterion called d-separation ('d' denotes directional), which serves to keep messages from spreading to irrelevant areas of the knowledge base. If a set of nodes Z d-separates set X from set Y, then the variables in X are guaranteed to be irrelevant to those in Y, once we know those in Z. The criterion is similar to ordinary separation in undirected graphs, except that paths traversing head-to-head arrows are discounted whenever these arrows have no descendants in the separating set Z.

Pearl, J., *Probabilistic Reasoning in Intelligent Systems: Networks of Plausible Inference*, Morgan Kaufmann, San Mateo, CA, 1988.

Judea Pearl

19 Beam Search

This is a search method in which heuristics are used to prune the Search Space[260] to a small number of nearly optimal alternatives. This set comprises the 'beam', and its members are then searched in parallel. Applications include speech recognition, vision, constraint directed search, and learning.

Oleinick, P., *The Implementation and Evaluation of Parallel Algorithms on C.mmp (PhD thesis)*, Carnegie Mellon University, Computer Science Department, 1978.

Helen Lowe

20 Behaviour Based Systems

Behaviour Based Systems are characterised by the way in which they decompose a problem. They have been used mostly for the control of robotics but are being increasingly applied to other fields. In the field of robotics a behaviour based system is one which decomposes the problem of control in terms of the overall behaviour of the system. The idea is that instead of having a large monolithic control structure (Model-based Systems[166]) control is split into a number of small parallel *behaviours* each of which accomplishes some small component of the overall task. Each behaviour on its own is not usually considered to be intelligent but rather the intelligence is considered to be an *emergent* property of the system as a whole.

Brooks, R.A., *A Robot that Walks; Emergent Behaviours from a Carefully Evolved Network*, AI Memo 1091 (1989), Artificial Intelligence Laboratory, Massachusetts Institute of Technology, Cambridge, MA.

Malcolm, C., Smithers, T. and Hallam, J., *An Emerging Paradigm in Robot Architecture*, Research Paper 447 (1989), Department of Artificial Intelligence, University of Edinburgh.

Martin Westhead

21 Bidirectional Search

In bidirectional search of a State Space[260], Backwards Search[16] and Forwards Search[96] are carried out concurrently. The program terminates when a common state is reached, since this means that a path has been found from the initial state to the goal state.

Barr, A. and Feigenbaum, E.A. (eds.), *The Handbook of Artificial Intelligence*, vol. 1, William Kaufmann, Los Altos, CA, 1981.

Korf, R.E., *Search: a survey of recent results*, Exploring Artificial Intelligence (Survey Talks from the National Conferences on Artificial Intelligence) (Shrobe, H.E., ed.), Morgan Kaufmann, San Mateo, CA, 1988, pp. 197–237 (Chapter 6).

Maarten van Someren

22 Binary and Grey Scale Moments

In computer vision, moments are scalar values that encode some property of the shape or distribution of an object. They are often used as a compact representation of simple binary shapes, because: (i) they are easy to compute (i.e., real-time hardware to compute moments exists), and (ii) a few different moments are often sufficient to characterise uniquely a limited set of shapes. It is also possible to calculate functions of the moments that are invariant to translation, rotation and scale changes.

The ij^{th} moment (m_{ij}) of a binary image about the centre of mass (x_0, y_0) is given by:

$$m_{ij} = \sum (x_0 - x)^i (y_0 - y)^j .$$

By varying the parameters i and j the resulting moment can yield a number of useful results such as the area and the eccentricity. For a grey scale image whose intensity can be expressed as a function, $f(x,y)$ of x and y, the moment m_{ij} is given by:

$$m_{ij} = \sum f(x,y)(x_0 - x)^i (y_0 - y)^j.$$

Gonzalez, R.C. and Wintz, P., *Digital Image Processing*, Addison-Wesley, Reading, MA, 1987 (second edition).

H.W. Hughes

23 Blackboard

An architectural technique developed principally for continuous reasoning 'real-time' systems. Based on the concept of 'experts' sitting round a blackboard, it attempts to co-ordinate the activities of a number of different Knowledge Sources (KS) by providing a global database between them, to which partial solutions to the problem under examination are posted.

The blackboard is divided into a number of abstraction levels, each one containing hypotheses for partial solutions linked to other levels by logical relationships; a monitor controls access to these hypotheses and inspects any changes in order to notify KSs of those of interest.

Each KS is independent of all others and interfaces externally in a uniform way that is identical across KSs and in which no KS knows which or how many others exist. In general, a KS monitors a level of the blackboard for conditions for which its knowledge is applicable, it then proposes to process those conditions by placing an item on the Agenda[7]. The agenda is a list of possible processing events from which the scheduler must choose the one most likely to lead to a complete problem solution. This decision making process can be controlled by a static set of rules about problem solving and a dynamic goal structure, which change as the solution progresses, to focus attention in a data-directed fashion. The chosen event is passed for execution to its instantiating KS.

The blackboard idea has also been extended to a hierarchical model, with differing concepts on different blackboards and knowledge sources 'piping' information between blackboards in the form of expectations, supports, refutations, etc.

Erman, L.D., Hayes-Roth, F., Lesser, V.R. and Reddy, D.R., *The Hearsay II speech understanding system: integrating knowledge to resolve uncertainty*, Computing Surveys **12** (1980), also in Blackboard Systems (Engelmore, R., and Morgan, T., eds.), Addison-Wesley, Wokingham, 1988, pp. 31–86.

Erman, L.D. and Lesser, V.R., *A multi-level organisation for problem solving using many, diverse, cooperating sources of knowledge*, Proceedings of IJCAI-75, pp. 483–490.

Martin Bennet, John Lumley

24 Bottom-up Parsing

(*Data-driven Parsing*)

In trying to parse a string with a grammar, if one starts with the string and tries to fit it to the grammar, this is bottom-up or data-driven parsing. For instance with a Context-free Grammar[54], one starts with a token in the string and works up from there on the basis of rules in the grammar which have that token on their right-hand side, trying to reduce eventually to the initial symbol.

Gazdar, G. and Mellish, C., *Natural Language Processing in LISP/PROLOG/POP-11: An Introduction to Computational Linguistics*, Addison-Wesley, Reading, MA, 1989, Section 5.2.

Henry Thompson

25 Boundary Detection

(Line Finding, Curve Detection)

The conventional approach to Image Segmentation[127] in early visual processing is Edge Detection[85] followed by boundary detection. Edge detection produces primitive edge elements, possibly with properties of magnitude and/or direction, at places in the image where the edge detection operator has 'fired'. The task of the boundary detection process is to produce a set of boundaries by appropriately connecting up primitive edge elements. There are two main methods:

- boundary tracking, and
- the generalised Hough Transform[124].

Since a boundary is, by definition, a connected set of edge elements, boundary tracking chooses any edge element and looks for its neighbours, which are then connected to the initial element to form a boundary. This search process is then continually repeated for the elements at the ends of the boundary until some termination criterion is met (such as not being able to find any nearby neighbours). In the Hough technique, each primitive edge element is transformed into a curve in a transform space representing all the possible boundaries the edge element could be part of. The transform-space curves produced by edge elements on the same image boundary should all intersect at the same point in transform space, which can be subjected to the inverse transform to characterise the boundary in the image. Originally applicable to straight boundaries only, this method has been generalised to deal with any curve describable in an analytic or tabular form.

Ballard, D.H. and Brown, C.M., *Computer Vision*, Prentice-Hall, Englewood Cliffs, NJ, 1982 (Chapter 5).

Bob Beattie

26 Branch-and-bound Algorithms

A solution technique for discrete optimisation problems which is widely used outside AI and is closely related to the A* Algorithm[2]. The task is to find the optimally valued tip of a walkable search tree. A subtree of the search tree need not be searched if a computation at its root yields a bound for its set of tip values which implies that none of them can be optimal.

Aho, A.V., Hopcroft, J.E. and Ullman, J.D., *Data Structures and Algorithms*, Addison-Wesley, Reading, MA, 1983, pp. 330–336.

Jim Doran

27 Breadth-first Parsing

Amounts to Breadth-first Search[28] of the tree of alternatives which arise in the parsing process. The opposite of Depth-first Parsing[70].

Gazdar, G. and Mellish, C., *Natural Language Processing in LISP/PROLOG/POP-11: An Introduction To Computational Linguistics*, Addison-Wesley, 1989.

Henry Thompson

28 Breadth-first Search

An uninformed graph searching strategy in which each level is searched before going to the next deeper level. This strategy is guaranteed to terminate with the shortest possible path to the goal node, if such a path exists, since any path to the solution that is of length n will be found when the search reaches depth n, and this is guaranteed to be before any node of depth greater than n is searched. This property is called admissibility. Breadth-first search is asymptotically optimal in time (see entry for Iterative Deepening[138]).

Nilsson, N.J., *Principles of Artificial Intelligence*, Tioga Pub. Co., Palo Alto, CA, 1980, and Springer-Verlag, Berlin, 1982.

Korf, R.E., *Search: a survey of recent results*, Exploring Artificial Intelligence (Survey Talks from the National Conferences on Artificial Intelligence) (Shrobe, H.E., ed.), Morgan Kaufmann, San Mateo, CA, 1988, pp. 197–237 (Chapter 6).

Dave Plummer

29 Caching

(*Infinite Recursion, Common Subgoals*)

When searching a goal tree, an unbounded amount of work can be wasted solving the same problem over and over. Even worse, an unbounded amount of work can be wasted in an infinite recursion. The standard technique of caching previous results can solve both problems. The cache is indexed by goal, normally implemented as a hash table. Each goal is associated with its current solution status: either it is solved (in which case the answer is stored), it has failed, or it is still open. Whenever a goal is to be proved, the cache is searched first. If it is not there, a cache entry marked open should be created. If the cached goal has succeeded or failed just use the stored answer. In the special case of Depth-first Search[71], infinite recursion occurs when an open cached goal is re-encountered; in this situation, the goal should be treated as though it has failed, though it should not be recorded as such in the cache since an alternative path might still succeed. It also worthwhile canonicalising the cache entries so that similar goals can be recognised. Earley's Algorithm[84] is an example of the use of caching. See also Case-based Reasoning[32].

Cohen, D. and Mostow, J., *Automatic program speedup by deciding what to cache*, Proceedings of IJCAI-85, 1985, pp. 165–172.

Don Cohen

30 Camera Calibration

A technique used to estimate the parameters of the geometric relationship between an image pixel's coordinates and the straight line in the external world along which light travelled to impinge on that pixel. The coefficients depend on both geometry (camera position and orientation) and optics (focal length, image-to-length scale factors, optical centre and possible distortions). The standard calibration model assumes linearity (i.e., no distortions from a pin-hole model). Under such an assumption, any visible external point with known coordinates determines one of a system of equations describing the relationship of the camera to the world. With enough reference points, the system can be solved for the unknown camera calibration parameters.

Tsai, R.Y., *Synopsis of recent progress on camera calibration for 3-D machine vision*, Autonomous Mobile Robots: Perception, Mapping, and Navigation **1** (1991), 191–200.

M.J. Orr

31 Cascaded Augmented Transition Network

(*CATN*)

An extension of Augmented Transition Network[13] parsing to use a sequence of ordinary ATNs that include among the actions on their arcs a special operation to transmit an element to the next ATN 'machine' in the sequence, with the first machine in the cascade taking its input from the input string. Provides a framework for separating, for example, syntactic, semantic and discourse tracking specifications into a cascade of ATNs, one (or more) for each domain. This clean separation of levels of processing means that different higher level partial hypotheses can share the same lower level processing, which in turn can be more flexible due to the loose coupling between the ATN machines (cf. coupling between subnetworks in an ordinary ATN).

Woods, W.A., *Cascaded ATN grammars*, American Journal of Computational Linguistics **6** (1980), 1–12.

John Carroll

32 Case-Based Reasoning

(*CBR*)

Case-based reasoning (CBR) is a problem solving paradigm which utilises the specific knowledge of previously experienced, concrete problem situations (cases). A new problem is solved by finding a similar past case, and reusing its solution in the new problem situation. CBR is a cyclic and integrated process of solving a problem and learning from this experience—central tasks of a CBR system are:

1. identify the current problem situation (case),
2. retrieve the most similar stored case (or cases),

3. reuse the information and knowledge in the retrieved case(s) to solve the new problem—this often involves adapting the old solution to fit the new situation,
4. revise and evaluate the proposed solution, and
5. retain the parts of this experience likely to be useful for future problem solving.

Case-based reasoning is claimed to be psychologically plausible, since humans appear to rely heavily on the use of past cases; according to some, expertise is like a library of past experience. CBR is also an approach to incremental, sustained learning, since a new experience is retained each time a problem has been solved, making it immediately available for future problems. When an attempt to solve the current problem fails, the reason for the failure is identified and remembered in order to avoid the same mistake in the future. Case-based reasoning can be considered as a form of Analogical Problem Solving[9] where only information from within the domain is used. CBR systems are restricted to variations on known situations and produce approximate answers but in large domains they will be faster than Rule-based Systems[213], producing solutions grounded in actual experience. See also Caching[29].

Riesbeck, C.K. and Schank, R.C., *Inside Case-based Reasoning*, Lawrence Erlbaum Associates, Hillsdale, NJ, 1989.

Kolodner, J.L., *Case-based Reasoning*, Morgan Kaufmann, San Mateo, CA, 1993.

Nicolas Nicolov

33 Case Frames

A widely used device for the determination and representation of text meaning, based on the organisation of information primarily around verbs, or actions, by case roles, e.g., *agent*, *instrument*, *location*. Case frames are usually small-scale structures oriented towards linguistic units like sentences, as opposed to the typical large-scale structures oriented to world knowledge represented by Frames[98]. Though reference to Fillmore's linguistically-motivated ideas is conventional, there is great variation in the treatment of every aspect of case frames, e.g., their relation to features of the surface text (is the subject of the sentence the agent?), their number (ten or thirty?), their status (obligatory or optional?), the constraints on their fillers (is the agent HUMAN?), etc. Conceptual Dependency[45], for example, uses a small number of deep cases referring primarily to underlying rather than surface relations. See also Case Grammar[34].

Bruce, B., *Case systems for natural language*, Artificial Intelligence **6** (1975), 327–360.

Karen Sparck Jones

34 Case Grammar

A model of grammar closely related to transformational grammar and originating from work described by Fillmore. Case Grammar attempts to characterise the set of semantic roles, called *cases*, that a noun phrase may fill. In Fillmore's paper, six cases were introduced, and each noun phrase must fill exactly one. Fillmore's cases are *agentive, instrumental, dative, factive, locative and objective.*

The main notion that Case Grammar attempts to capture is that in comparing sentences like *John read the book* and *This book reads easily* the subject of the first sentence (*John*) is the agent (John is reading the book), whereas the subject of the second (*This book*) is the object (the book is being read).

Fillmore, C.J., *The case for case*, Universals in Linguistic Theory (Bach, E. and Harms, R., eds.), Holt, Rinehart and Winston, New York, 1968, pp. 1–88.

Samlowski, W., *Case grammar*, Computational Semantics: An Introduction to Artificial Intelligence and Natural Language Comprehension (Charniak, E. and Wilks, Y., eds.), North-Holland, Amsterdam, 1976, pp. 55–72.

Allen, J., *Natural Language Understanding*, Benjamin, Menlo Park, CA, 1987.

John Beavan

35 Categorial Grammar

A grammatical formalism, first developed by Adjukiewicz, and later resurrected by Bar-Hillel, Steedman and others, in which grammatical syntactic categories (such as *nouns, adjectives*) are regarded as *functors* over certain *arguments.* Unlike in phrase structure grammars, most of the combinatory properties of the words are encoded in the lexical entries, and the grammar then has a small number of rules. The main one is *function application* which puts a functor and its argument together, but others, such as *type-raising*, and *function composition* may also be used. One of its variants is combinatory categorial grammar (CCG), which attempts to relate the grammar rules to Curry's combinatory logic.

Ades, A. and Steedman, M.J., *On the order of words*, Linguistics and Philosophy **4** (1982), 517–518.

Uszkoreit, H., *Categorial unification grammar*, Proceedings of COLLING-86, 1986, pp. 187–194.

Oehrle R.T., Bach E. and Wheeler D. (eds.), *Categorial Grammars and Natural Language Structures*, D. Reidel Publishing Co., Dordrecht, 1988.

John Beavan

36 Cellular Arrays

A class of parallel computers consisting of an array of small processors each executing the same instruction on its local data. The processors are often one-bit processors (a bit-serial array) and are often arranged in a regular lattice using hexagonal or square tessellation (with or without diagonals). Such arrays are particularly useful for simple image processing where each image pixel is generally assigned to a separate processor. Typical examples are DAP (from ICL), CLIP (from University College London), MPP (from Goodyear Aerospace) and GRID (from GEC). See Propagation in Cellular Arrays[218].

Danielsson, P.E. and Levialdi, S., *Computer architectures for pictorial information systems*, Computer **14** (1981), 53–67.

Dave Reynolds

37 Certainty Factors

Certainty factors are used in Production Rule Systems[213], such as MYCIN, to express and propagate degrees of belief. They are numbers in the range $[-1, +1]$. $+1$ indicates certainty; -1 indicates impossibility. Measures of belief and disbelief are collected separately and combined by subtracting the latter from the former. Certainty factors are updated to take into account fresh evidence, the laws of combination differing from Bayesian Inference[17] in that they are truth functional—the certainty of a formula is a unique function of the certainty of its subformulae. Approximate probabilistic interpretation of certainty factors can be given in terms of the likelihood ratio:

$$L = \frac{P(e|h)}{P(e|\neg h)} ,$$

and the transformation:

$$CF = \frac{(L-1)}{(L+1)} .$$

The certainty factor propagation method would produce coherent belief updates if no two rules emanate from the same premise.

Heckerman, D., *Probabilistic interpretations for MYCIN's certainty factors*, Uncertainty in Artificial Intelligence (Kanal, L.N., and Lemmer, J.F., eds.), North-Holland, Amsterdam, 1986, pp. 167–196.

Helen Lowe

38 Chart Parsing

Chart parsing is an approach to non-deterministic parsing developed by Kay and Kaplan based on earlier work by Earley, Kay and Colmerauer. In contrast to that earlier work, in which the *chart* was a (in some cases enriched) well-formed substring table for recording intermediate results, the later systems use the chart as the active agent in the parsing process.

The chart is a directed graph, with two sorts of edges—active and inactive. Inactive edges record the existence of complete constituents. Active edges record hypothesised *in*complete constituents. The parsing process itself consists of adding new edges in response to the meeting of active with inactive edges. As a record of an incomplete constituent, an active edge must carry some indication of how it may be extended, e.g., a dotted context-free rule or a state in a network grammar, such as Recursive Transition Network (RTN) or ATN[13]. When an active edge and an inactive edge meet for the first time, if the inactive edge satisfies the active edge's conditions for extension, then a new edge will be constructed.

If initial hypotheses about constituents are keyed by active edges and their needs, parsing will be top-down (see Top-down Parsing[279]). If on the other hand these hypotheses are keyed by inactive edges, parsing will be bottom-up (see Bottom-up Parsing[24]). One of the principal advantages of the chart parsing methodology is that it easily supports a variety of grammatical formalisms, rule invocation strategies, and control regimes. See (Thompson and Ritchie 1984) for an elementary introduction, and (Kay 1986) for a detailed theoretical analysis.

Thompson, H.S. and Ritchie, G.D., *Natural language processing*, Artificial Intelligence: Tools, Techniques, and Applications (O'Shea, T. and Eisenstadt, M., eds.), Harper and Row, New York, 1984, pp. 358–388.

Kay, M., *Algorithm schemata and data structures in syntactic processing*, Readings in Natural Language Processing (Grosz, B.J., Sparck Jones, K. and Webber, B.L., eds.), Morgan Kaufmann, Los Altos, CA, 1986, pp. 35–70.

Henry Thompson

39 Circumscription

Circumscription captures the idea of 'jumping to conclusions', namely that the objects that can be shown to have a certain property are all the objects that satisfy that property. For example, in the missionaries and cannibals problem, we are told that three missionaries and three cannibals want to cross the river in a two-person boat such that the cannibals never outnumber the missionaries in the boat or on either bank. We assume that there are no more cannibals around; the three mentioned in the problem are all there are. More generally, we conjecture that the tuples $\langle x, y, ..., z\rangle$ that can be shown to satisfy a relation $P(x, y, ..., z)$ are all the tuples satisfying this relation. Thus we *circumscribe* the set of all relevant tuples.

Circumscription is a formalised rule of conjecture. Domain circumscription (also known as minimal inference) conjectures that the known entities are all there are. Predicate circumscription assumes that entities satisfy a given predicate only if they have to on the basis of a collection of known facts; since this collection can be added to subsequently. Circumscription together with first order logic allows a form of Non-monotonic Reasoning[182]. Suppose $A(P)$ is a sentence of first order logic containing a predicate symbol $P(\underline{x})$, where $\underline{x} = x_1, x_2, \ldots, x_n$, and that $A(\Phi)$ is the result of replacing all occurrences of P in A by the predicate expression Φ. Then the predicate *circumscription* of P in $A(P)$ is the schema:

$$A(\Phi) \wedge \forall\underline{x}(\Phi(\underline{x}) \rightarrow P(\underline{x})) \rightarrow \forall\underline{x}.(P(\underline{x}) \rightarrow \Phi(\underline{x})) \; .$$

The sentences that follow from the predicate circumscription of a theory are those which are true in all the minimal models of that theory.

McCarthy, J., *Circumscription—a form of non-monotonic reasoning*, Artificial Intelligence **13** (1980), 27–39.

Helen Lowe

40 Classification

(*ID3*)

A Discrimination Net[76] can be built from a set of classified items. The set is repeatedly split into subsets by the value of a predicate, until each subset contains members of one class only. The series of splits defines the discrimination net.

To increase the efficiency of the resulting discrimination net, the predicates on which the items are split are chosen using information theoretic techniques so that they make the number of instances in each subset as similar as possible. This technique has been successfully applied to the problem of classifying chess positions as won or drawn.

Quinlan, J.R., *Discovering rules by induction from large collections of examples*, Expert Systems in the Micro-electronic Age (Michie, D., ed.), Edinburgh University Press, Edinburgh, 1979, pp. 168–201.

Quinlan, J.R., *Induction of decision trees*, Machine Learning **1** (1986), 81–106.

Maarten van Someren

41 Clausal Form

A normal form for Predicate Calculus[208] formulae borrowed from mathematical logic, and much used in Automatic Theorem Proving[276]. The clausal form of a formula is produced by first converting it into prenex normal form, Skolemising[255], and converting the result into conjunctive normal form. The final formula has a model if and only if the original formula does. A formula in clausal form consists of a conjunction of clauses. Each clause is a disjunction of literals. Each literal is either an atomic sentence or the negation of an atomic sentence, where an atomic sentence is a predicate applied to some terms.

Horn clauses are a very important subclass of clausal form formulas. They have at most one un-negated literal. Using the Kowalski form (where $\neg P_1 \vee \ldots \vee \neg P_m \vee Q_1 \vee \ldots \vee Q_n$ is written as $P_1 \& \ldots \& P_m \rightarrow Q_1 \vee \ldots \vee Q_n$) they are formulas of the form:

$$\mathrm{P}_1 \;\&\; \mathrm{P}_2 \;\&\; \ldots \;\&\; \mathrm{P}_n \;\rightarrow\; \mathrm{Q} \qquad \text{or} \qquad \mathrm{P}_1 \;\&\; \mathrm{P}_2 \;\&\; \ldots \;\&\; \mathrm{P}_n \;\rightarrow$$

where each of the P_i and Q are atomic sentences.

Horn clauses have several important properties when viewed from mathematical logic. In addition, they form the basis for the logic programming language Prolog[216]: each predicate in a Prolog program has a Horn clause definition. The above formulae would be written:

$$Q :- \mathrm{P1}, \mathrm{P2}, \ldots, \mathrm{Pn}. \qquad \text{and} \qquad ?- \mathrm{P1}, \mathrm{P2}, \ldots, \mathrm{Pn}.$$

respectively as Prolog programs.

Chang, C.L. and Lee, R.C.T., *Symbolic Logic and Mechanical Theorem Proving*, Academic Press, New York, 1973.

Kowalski, R.A., *Logic for Problem Solving*, Elsevier North-Holland, New York, 1979.

Chang, C.C. and Keisler, H.J., *Model Theory*, Elsevier North-Holland, New York, 1973.

Alan Bundy, Martin Merry

42 Competitive Networks

Self-organising (or Unsupervised Learning[289]) algorithms train a neural network to find patterns, categories or correlations in data without a feedback signal from a teacher. In competitive networks, this information is represented by a single output unit or class of units. The technique is useful for data encoding and compression, combinatorial optimisation (e.g., Travelling Salesman Problem), function approximation, image processing and statistical analysis. Simple competitive networks contain one layer of output units, each fully connected to a layer of (N-dimensional) inputs by excitatory connections. The response of an output is simply the input signal amplified (or diminished) in proportion to the strength of the weighted connections along which the input signal is carried. The output node most excited by an input is said to 'win' and, as a reward, gets the efficacy of its weighted connections adjusted so that it encodes a stronger description of the current input. (Self-Organising Feature Mapping[240] algorithms are a special case of competitive learning where the geometrical arrangement of output units (or groups of units) preserves topological information about the input space.) There is no guarantee that competitive learning will converge to the best solution. Some stability can be imposed by decreasing weight changes over time—thus freezing the learned categories, but this affects the network's plasticity (its ability to react to new data). Carpenter and Grossberg's Adaptive Resonance Theory seeks a solution to the stability-plasticity dilemma.

Hertz, J., Krough, A. and Palmer, R.G., *Introduction to the Theory of Neural Computation*, Addison-Wesley, Redwood City, CA, 1991.

Carpenter, G.A. and Grossberg, S., *A massively parallel architecture for a self-organising neural pattern recognition machine*, Computer Vision, Graphics and Image Processing **37**, 54–115.

Ashley Walker

43 Complexity Measures

A complexity measure measures some aspect of the complexity of the current problem state in a search problem. For instance, the depth of a goal is the length of the path from the current goal to the origin of the Search Space[260]. Complexity measures are sometimes associated with the labels of nodes in a search space, especially when these are logical expressions describing the current goal; e.g., the depth of function nesting of an expression is the maximum amount of nesting in the functions in it. The size of an expression is the number of symbols in it. These symbols can also be weighted and the weights totalled. Complexity measures can be used for pruning, in which case they are usually called bounds. In pruning, parts of the search space whose measure exceeds a threshold are not searched. They can also be used as components in the evaluation function of heuristic search.

Nilsson, N.J., *Principles of Artificial Intelligence*, Tioga Pub. Co., Palo Alto, CA, 1980, and Springer-Verlag, Berlin, 1982.

Alan Bundy

44 Conceptual Clustering

Conceptual clustering is an Unsupervised Learning[289] procedure which builds a Discrimination Net[76] given a set of input patterns. It begins with the root node of a tree structure, all of the input patterns being attached to that node. A decision is made as to how split the set of all input patterns into two (or more) more specific classes. Each of these classes is represented by sub-nodes in the tree structure, each input pattern being attached to the node representing its class. The splitting of the tree continues until all nodes are singleton sets. Different approaches to conceptual clustering use different methods to decide how best to split the tree. COBWEB chooses the partition of a set of input patterns which maximises a measure of the total intra-class similarity and inter-class dissimilarity. Intra-class similarity can be expressed as the conditional probability of an object possessing a particular property given that the object belongs to a particular class. Inter-class dissimilarity can be expressed as the conditional probability of a particular class given a property. This is summed over all classes and their members for each possible partition of the set of input patterns.

Conceptual clustering is related to the statistical technique of cluster analysis. The principal difference is that while cluster analysis requires that all the data (input patterns) are present at the beginning of process, conceptual clustering works incrementally, i.e., it adjusts its current partitioning of the space of input patterns as each new pattern is presented. This has the consequence that conceptual clustering methods may converge to suboptimal partitions. COBWEB attempts to overcome this problem by employing node merging and splitting operators. These effectively allow the learner to backtrack within the search space of possible partitions.

Fisher, D. and Langley, P., *Approaches to conceptual clustering*, Proceedings of IJCAI-85, Morgan Kaufmann, Los Altos, CA, 1985, pp. 691–697.

Jeremy Wyatt

45 Conceptual Dependency

A theory of meaning representation developed by Roger Schank and extensively exploited at Yale, relying on deep 'conceptual' Semantic Primitives[245] and Case Frames[33] and providing a strong decomposition of word and text meaning. The emphasis on key primitive acts and required properties of the fillers of their (obligatory) roles is used to drive primarily semantic expectation-based parsing.

Schank R.C. (ed.), *Conceptual Information Processing*, North-Holland, Amsterdam, 1975.

Schank R.C. and Riesbeck, C.K. (eds.), *Inside Computer Understanding: Five Programs Plus Miniatures*, Lawrence Erlbaum Associates, Hillsdale, NJ, 1981.

Karen Sparck Jones

46 Conceptual Graphs

(*CGs*)

Conceptual Graphs (CGs) are a knowledge representation formalism, a variant of Semantic Networks[244]. CGs were developed by John Sowa and stem from the Existential Graphs of Charles Sanders Peirce which are a graphic notation for Classical Logic[208]. A conceptual graph is a structure of concepts and conceptual relations linked with arcs where every arc connects a conceptual relation to a concept. The types of the concepts and the relations form concept- and relation-hierarchies (lattices). A graph can be embedded in another graph by means of the context mechanism. The default existential quantifier and the possibility to negate contexts give CGs an equivalent power to classical predicate calculus. CGs have definitional mechanisms that can support extensions to the core in a controlled, systematic way. Because CGs can be mapped to classical predicate calculus (or order-sorted logic), they are thus seen as a (graphical) notation for logic. However, it is the topological nature of formulas which CGs make clear, and which can be exploited in reasoning and processing. CGs are intuitive because they allow humans to exploit their powerful pattern matching abilities to a larger extent than does the classical notation—thus, CGs are particularly useful for the interchange of knowledge between humans and computers. CGs can be viewed as an attempt to build a unified modelling language and reasoning tool. They can model data, functional and dynamic aspects of systems. They form a unified diagrammatic tool which can integrate entity-relationship diagrams, finite-state machines, Petri nets, and dataflow diagrams. CGs have found wide application in systems for information retrieval, database design, expert systems, conceptual modelling and natural language processing. See Graph Matching[112].

Sowa, J., *Conceptual graphs summary*, Conceptual Structures: Current Research and Practice (Nagle, T.E., Nagle, J.A., Gerholz, L.L. and Eklund, P.W., eds.), Ellis Horwood, Chichester, 1992, pp. 3–51.

Nicolas Nicolov

47 Connection Calculus

(*Clause Graph Resolution, Connection Graph*)

The connection calculus is a proof system for first-order classical Predicate Calculus[208] based on the notion of a matrix. The method is due to Wolfgang Bibel (1981) and grew out of an analysis of more standard sequent and tableau methods. A similar method of 'matings' was developed independently by Peter Andrews in 1981.

A formula or entailment of the first-order predicate calculus can be displayed as a (nested) two-dimensional matrix. For example, the entailment $A \rightarrow B, B \rightarrow C \vdash A \rightarrow C$, is represented as the matrix:

$$\begin{pmatrix} A \\ \overline{B} \end{pmatrix} \quad \begin{pmatrix} B \\ \overline{C} \end{pmatrix} \quad (\overline{A} \quad C)$$

where $\overline{A}$ indicates that the occurrence of A is negative, i.e., inside an odd number of explicit or implicit negation signs. The matrix is made up of three columns, the first two of which themselves are matrices with one column and two rows. The third matrix has two columns and one row. Ignoring the brackets and reading horizontal separation as conjunction and vertical separation as disjunction we have the conjunctive normal form of the original entailment. (We could just as easily work with the disjunctive normal form by attempting to refute the entailment rather than prove it.) The paths through this matrix are lists of literals that contain exactly one element from each column of the matrix, and a connection is a pair of literals on a path that are complementary (i.e., identical atoms but complements with respect to negation). In our example, $A, \overline{C}, \overline{A}$ is one path, $A, \overline{C}, C$ another. $A, \overline{A}$ and $C, \overline{C}$ are connections on these paths.

The basic characterisation of consequence (or validity) on which the connection calculus rests is that an entailment $\Gamma \vdash A$ holds (i.e., A is a logical consequence of the formulae of Γ) if and only if every path through the matrix representation of the entailment contains a complementary connection. This is a version of Gentzen's Hauptsatz for the logic. In our example, the connections $A, \overline{A}$, $B, \overline{B}$ and $C, \overline{C}$ are said to span the matrix because every path through it contains a connection from this (three element) set of connections. The entailment therefore holds. Various so-called path-checking algorithms have been designed that simulate resolution strategies, and to take extra advantage of the form of the matrix (see Bibel 1982). In the presence of quantifiers complementarity of connections is calculated by a unification algorithm as in standard Resolution[234]. The method extends to classical type theory (see Andrews' work) and non-classical logics (Wallen's 1990 book).

Andrews, P.B., *Theorem proving via general matings*, JACM **28** (1981), no. 2, 193–214.

Bibel, W., *On matrices with connections*, JACM **28** (1981), no. 4, 633–645.

Bibel, W., *Automated Theorem Proving*, Vieweg, Braunschweig Wiesbaden, 1982 (revised 1987).

Wallen, L.A., *Automated Proof Search in Non-classical Logics*, MIT Press, Cambridge, MA, 1990.

Lincoln Wallen

48 Connectionism

(*Neural Networks, Neuro-computation, Parallel Distributed Processing*)

Connectionism is an alternative computing paradigm to that offered by von Neumann. It is inspired by ideas from neuroscience and draws much of its methodology from statistical physics. Connectionist architectures exploit the biologically proven principle that sophisticated computation can arise from a network of appropriately connected simple processing units (e.g., neurons) which receive simple excitatory and inhibitory messages from other units and perform some function on these inputs in order to calculate their output. Some of the earliest work in the field is due to McCulloch and Pitts who proposed a simple model of a neuron as a binary threshold unit. They proved that a synchronous assembly of these artificial neurons are capable, in principle, of universal computation.

The operation of a particular network depends upon the pattern of interconnectivity (i.e., the 'synaptic strength' or 'weights') between elemental units. Although these connections may be programmed directly from model equations, much of the utility of connectionist networks lies in their ability to learn the connections necessary to solve a problem from presentation of example data. Moreover, networks can generalise over examples so as to interpolate and extrapolate new relationships between members of the data which are not explicitly represented.

Learning in connectionist networks may be guided by a supervisor or teacher—in which case a direct comparison is made between network response to input data and a known correct response. (Supervised Learning[250] involves a specialised case of reinforcement learning, where the content of the feedback signal only describes whether each output is correct or incorrect.) However, if the learning goal is not defined, Unsupervised Learning[289] algorithms can be used to discover categories through the correlations of patterns in the input data.

McCulloch, W.S. and Pitts, W., *A logical calculus of ideas immanent in nervous activity*, Neurocomputing: Foundations of Research (Anderson, J.A. and Rosenfeld, E., eds.), MIT Press, Cambridge, MA, 1988, pp. 15–42.

Rumelhart, D.E., McClelland, J.L. and The PDP Research Group (eds.), *Parallel Distributed Processing: Explorations in the Microstructure of Cognition. Volume 1*, MIT Press, Cambridge, MA, 1989.

Hertz, J., Krough, A. and Palmer, R.G., *Introduction to the Theory of Neural Computation*, Addison-Wesley, Redwood City, CA, 1991.

Ashley Walker

49 Constituent Likelihood Grammar

Natural language analysis techniques based on generative grammars normally make a sharp distinction between grammatical and ungrammatical sequences and thus fail when exposed to authentic discourse. Probabilistic techniques for (i) syntactic category assignment and (ii) syntactic structure determination use only the relative frequency of constructions and so do not rule anything out as ungrammatical. The CLAWS system developed at Lancaster University for word tagging using tag pair transition frequencies derived from a large, heterogeneous text corpus achieves 95-96% accuracy.

Atwell, E.S., *Constituent-likelihood grammar*, The Computational Analysis of English: A Corpus-based Approach (Garside, R., Sampson, G. and Leech, G., eds.), Longman, London, 1987, pp. 57–65.

Geoffrey Sampson, Karen Sparck Jones

50 Constraint Logic Programming

(CLP)

Constraint logic programming is a paradigm coined for solving combinatorial problems. It has been used successfully for a variety of practical applications from scheduling to financial analysis. Constraint handling has been introduced into Logic Programming[153] to simplify the expression of problems and to dramatically improve program efficiency. Essentially the constraints are used to prune the search tree defined by the logic program in which they are embedded.

Three different kinds of constraints have been introduced: basic constraints, propagation constraints and reactive constraints. During program execution, basic constraints are added to a global store of constraints and the system checks the consistency of the whole constraint store. Propagation constraints are used to restrict possible constraint stores: they can also be used actively to continually produce new information, as the constraint store grows. Reactive constraints exhibit a behaviour which is specified by rules, whose firing is driven by the constraint store.

An embedding of constraints in logic programming, and illustrations of its use, are presented in the article referenced below.

Van Hentenryck, P., Simonis, H. and Dincbas, M., *Constraint satisfaction using constraint logic programming*, Artificial Intelligence **58** (1992), 113–159.

Mark Wallace

51 Constraint Networks

Constraint networks offer a graphical representation of Constraint Satisfaction[52] problems that can be used to control search efficiently. A constraint network is a graph (or hypergraph) in which nodes represent variables and arcs represent pairs (or sets) of variables which are included in a common constraint. The topology of this network uncovers opportunities for problem decomposition techniques and provides estimates of the problem complexity prior to actually performing the search. The following are typical network-based techniques.

- Cycle Cutset is a method based on preferentially instantiating variables which intercept cycles in the constraint graph. Once all cycles are intercepted, the rest of the problem becomes tree-structured and can be solved in linear time. The complexity of this scheme is exponential in the size of the cutset found.
- Adaptive Consistency is a method based on enforcing local consistency with adjustable scope. Variables are called sequentially, and local consistency is enforced recursively among their uncalled neighbours. Each such operation would normally result in connecting the neighbours with extra links, called 'induced constraints'. When all variables are called, a backtrack-free solution can be found by instantiating the variables in the reverse order.
- Tree Clustering is a method of systematic regrouping constraints into hierarchical structures capable of supporting search without backtracking. The complexity of this transformation is exponential in the size of the largest cluster generated and its main advantage is that, once the clusters are structured, queries can be answered in linear time.

Other graph-based techniques include backjumping, an improvement on backtracking where, at a dead-end situation, the algorithm retracts to the most recent node connected to the dead-end variable (also see Dependency Directed Backtracking[69]), and decomposition to a tree of connected components where the connected components in the constraint graph are solved first and an overall solution is assembled using a linear tree algorithm.

Dechter, R., *Enhancement schemes for constraint processing: backjumping, learning, and cutset decomposition*, Artificial Intelligence **43** (1990), 273–312.

Dechter, R. and Pearl, J., *Tree clustering for constraint networks*, Artificial Intelligence **38** (1989), 353–366.

Freuder, E.C., *A sufficient condition for backtrack-bounded search*, JACM **32** (1985), no. 4, 755–761.

Rina Dechter

52 Constraint Satisfaction and Propagation

(*Consistent-labelling, Satisfaction Assignment, CSP*)

Constraint satisfaction is a class of problems in which knowledge is expressed declaratively as a collection of explicit, categorical constraints over a set of possibilities. These include, for example, scene labelling, scene matching, space planning problems, database consistency-maintenance, query-answering, graph isomorphism detection, the Graph Colouring problem, and many puzzles such as crypto-arithmetic.

The *Constraint-Satisfaction Problem* consists of a set of n variables, $V_1 \dots V_n$, each represented by its domain values (not necessarily numerical), $D_1 \dots D_n$, and a set of constraints. A constraint $C_i(V_{i_1}, \dots V_{i_j})$ is a subset of the Cartesian product $D_{i_1} \times \dots \times D_{i_j}$ that specifies which values of the variables are compatible with each other. A solution is an assignment of values to all the variables which simultaneously satisfy all the constraints, and the most common task associated with these problems is to find one or all solutions. In practice, the constraints may be specified either as truth-tables, or as lists of the satisfying tuples or, more concisely, in analytic form as expressions (such as $V + V^5 \leq 9$) in some 'constraint languages', algebraic or otherwise. See also the entry on Relaxation Labelling[233].

The common algorithm for solving CSPs is the backtrack search algorithm. In its primitive version, called Chronological Backtracking[71], the algorithm traverses the variables in a predetermined order, provisionally assigning consistent values to a subsequence $(V_1, \dots, V_i)$ of variables and attempting to append to it a new instantiation of V_{i+1} such that the whole set is consistent. If no consistent assignment can be found for the next variable V_{i+1}, a dead-end situation occurs, the algorithm 'backtracks' to the most recent variable, changes its assignment and continues from there. Chronological backtracking suffers from many maladies and several cures have been offered and analysed in the AI literature. These can be classified into *look-ahead schemes*, and *look-back schemes*. The former are concerned with the decision of what variable to instantiate (e.g., dynamic search rearrangement, max-cardinality search, minimum width), or what value to assign next among all the consistent choices available (e.g., forward-checking, full-lookahead, partial look-ahead, constraint propagation). Look-back schemes concern the decision of where and how to backtrack in case of a dead-end situation. The two fundamental ideas in look-back schemes are *go-back to source of failure*, and *constraint recording* or learning. The former detect and change decisions that caused the dead-end while skipping irrelevant decisions (e.g., selective backtracking, backjumping), while constraint-recording schemes identify the conflict-sets representing the reasons for the dead-end and record them as explicit constraints to disable the same conflicts from occurring in the future (e.g., Dependency Directed Backtracking[69]).

Kumar, V., *Algorithms for constraint-satisfaction problems: a survey*, AI Magazine **13** (1992), no. 1, 32–44.

Rina Dechter

53 Constructive Solid Geometry

(*CSG*)

Constructive solid geometry is the volumetric representation of objects using compositions of simple solids. Usually the solids are simple cuboids, cylinders, cones or spheres with an arbitrary scale. These undergo rigid transformations in space before being combined using the set operations. Complex shapes are constructed by combining the basic solids using the AND and OR operations while vacancies or holes are formed using the NOT operation to 'bore out' the required shape. These operations reflect the way in which the object being modelled was originally constructed. See also Generalised Cylinders[106] and Superquadrics[268].

Requicha, A.A.G. and Chan, S.C., *Representation of geometric features, tolerances and attributes in solid modellers based on constructive geometry*, IEEE Journal of Robotics and Automation **2** (1986), 156–166.

H.W. Hughes

54 Context-free Grammar

(*CFG, Context-free Phrase Structure Grammar*)

A context-free grammar is a collection of context-free phrase structure rules. Each such rule names a constituent type and specifies a possible expansion thereof. The standard notation is:

$$LHS \rightarrow RHS_1, \ldots, RHS_n.$$

where LHS names the constituent, and RHS_1 through RHS_n the expansion. Such rules are context-free rules because the expansion is unconditional—the environment of the constituent to be expanded is irrelevant.

A collection of such rules together with an initial symbol, usually S, is a context-free grammar. The constituents expanded (those appearing on the left-hand side) are called the non-terminals. Those not expanded (appearing only on the right-hand side) are called the terminals. There are two standard ways of interpreting such a grammar as specifying a language. The rewriting interpretation says that a grammar generates the set of strings of terminals which can be produced by the following non-deterministic method:

1. Write down the initial symbol.
2. Choose a non-terminal symbol in the string, and a rule from the grammar which expands it. Replace the chosen instance of the non-terminal with the expansion given in the rule.
3. If no non-terminals remain in the string, the process is complete. Otherwise, go back to step 2.

The well-formedness interpretation actually generates trees, not strings. It simply admits to the language all singly rooted trees whose root is the initial symbol and for each of whose non-leaf nodes there is a rule in the grammar such that LHS is the node label and RHS_n are the labels of its descendants in order.

Gazdar, G., *Phrase structure grammar*, The Nature of Syntactic Representations (Jacobson, P. and Pullum, G.K., eds.), D. Reidel Publishing Co., Dordrecht, 1981, pp. 131–186.

Henry Thompson

55 Context-sensitive Grammar

A grammar in the Chomsky hierarchy, similar to a Context-free Grammar[54], but in which the right-hand side consists of the left-hand side with a single symbol expanded. The languages characterised by such a grammar can be recognised/parsed deterministically using an amount of storage space proportional to the length of the input. There has been considerable debate as to where exactly between context-free and context-sensitive natural languages fit.

Winograd, T., *Language as a Cognitive Process, Volume 1: Syntax*, Addison-Wesley, Reading, MA, 1983, pp. 143–147.

John Beavan

56 Contour Generator

A contour generator is a set of points on a generating object surface that projects to an occluding contour in an image. A special case of a contour generator is an extremal boundary of an object, this is a boundary along which a surface turns smoothly away from the viewer. At extremal boundaries the surface normal is perpendicular to both the line of sight and the occluding contour projected into the image.

Barrow, H.G. and Tenenbaum, J.M., *Retrospective on 'interpreting line drawings as three-dimensional surfaces'*, Artificial Intelligence **59** (1993), 71–80.

Marr, D., *Analysis of occluding contour*, Proceedings of the Royal Society, London **B197** (1977), 441–475.

T.P. Pridmore, S.R. Pollard, S.T. Stenton

57 Contradiction Backtracing

A method to discover and correct faulty hypotheses in a theory. If a proposition that is derived from a set of hypotheses turns out to be false in terms of a model, the derivation can be used to identify the faulty hypotheses. Contradiction backtracing uses the trace of a Resolution[234] proof, in which the false proposition was the goal. The proof is traced backwards from the empty clause. Each clause is semantically evaluated by the user or in a standard model. If it is false, the negated parent clause is considered next, and otherwise the positive parent. The substitutions that allowed the resolution steps are accumulated and applied to each following clause. The procedure will finally lead to a hypothesis which is false.

Shapiro, E.Y., *An algorithm that infers theories from facts*, Proceedings of IJCAI-81, vol. 1, 1981, pp. 446–451.

Maarten van Someren

58 Contrast Sensitivity Function

The contrast sensitivity function is a normalised description of a system's sensitivity to spatial frequencies in terms of the contrast required to perform some perceptual task. For detection tasks, the human C.S.F. peaks at around 3–5 cycles/deg and reaches zero at about 60 cycles/deg. See Modulation Transfer Function[169].

Wilson, H.R. and Giese, S.C., *Threshold visibility of frequency gradient patterns*, Vision Research **17** (1977), 1177–1190.

T.P. Pridmore, S.R. Pollard, S.T. Stenton

59 Convolution

The application of a mathematical operation to each neighbourhood in an image is called convolution. The operation is defined by a 'mask' specifying for each neighbourhood, how many points it contains and how the corresponding image point affects the computations. Each location in the operator mask contains a weighting value, these are multiplied by the value of the corresponding image location and the results summed to give the convolution value for that neighbourhood. Doing this for all neighbourhoods produces a new array of values. Mathematically, the convolution integral is the integrated cross product of a weighting function with an image. See Local Grey-level Operations[151].

Frisby, J.P., *Seeing: Illusion, Brain, and Mind*, Oxford University Press, Oxford, 1979.

T.P. Pridmore, S.R. Pollard, S.T. Stenton

60 Curvature Maps

The representation of the curvatures of an image considered as a surface—usually applied to depth maps. The normal curvature of a surface at a point reaches its maximum and minimum, the principal curvatures, at directions in the surface which are perpendicular. At an umbilic point, e.g., on a plane or sphere, the normal curvature is independent of direction. The Gaussian curvature (frequently denoted K), the product of the principal curvatures, and the mean curvature (frequently denoted H), the mean of the principal curvatures, may be used to classify the surface type into perceptually significant categories according to their sign.

Besl, P. and Jain, R., *Intrinsic and extrinsic surface characteristics*, IEEE CS Conference on Computer Vision and Pattern Recognition, 1985, pp. 226–233.

R.M. Cameron-Jones

61 Data-directed Control

(*Data-driven Control, Bottom-up, Forward Chaining*)

A technique for interpretation or evaluation of a set of clauses which represent constraints or equations on unknown data items. The evaluation proceeds in pseudo-parallel Breadth-first Search[28] fashion starting with those predicates where enough data items have values. These compute values which are then used in other clauses which are added to a queue for evaluation. Evaluation proceeds until no more predicates can be evaluated. Evaluation of some predicates may generate other clauses, in which case care must be taken to avoid an explosion of partially evaluated clauses. Control depends not on the initial order of the clauses, but on the order in which data items get their value. Demons[67] can be used to implement data-directed control.

Elcock, E.W., McGregor, J.J. and Murray, A.M., *Data directed control and operating systems*, The Computer Journal **15** (1972), no. 2, 125–129.

P.M.D. Gray

62 Decision Theory

Decision theory provides a basis for making choices in the face of uncertainty, based on the assignment of probabilities and payoffs to all possible outcomes of each decision. The space of possible actions and states of the world is represented by a decision tree, whose nodes may be either:

- decision nodes (actions which may be performed by the user),
- chance nodes (events over which the user has no control, with associated probabilities), or
- terminal nodes (outcomes) to which utilities are attached.

The decision tree is evaluated by calculating the expectation of the utility at each chance node and applying the maximisation operator at each decision node. The decision tree can be generated automatically from more economical representations of the decision environment such as influence diagrams or Bayesian Networks[18]. Other methods for dealing with uncertainty are Bayesian Inference[17], Dempster-Shafer Theory[68] and Certainty Factors[37].

Pearl, J., *Probabilistic Reasoning in Intelligent Systems: Networks of Plausible Inference*, Morgan Kaufmann, San Mateo, CA, 1988.

Helen Lowe

63 Deductive Program Synthesis

(*Extracting Answers from Proofs*)

Program synthesis is the derivation of a program to meet a given specification. According to the deductive approach, program synthesis is regarded as a theorem-proving task. The system proves the existence of an output that satisfies the specification. This proof is restricted to be constructive in the sense that it must indicate a computational method of finding the output. This method becomes the basis for a program that is extracted from the proof.

Manna Z. and Waldinger R., *The Logical Basis for Computer Programming*, vol. 1, Addison-Wesley, Reading, MA, 1985.

Richard Waldinger

64 Default Logic

(*Default Reasoning*)

In reasoning about incompletely specified worlds, it may be necessary to make default assumptions that lead to tentative conclusions. For example, if we know that Tweety is a bird, we may assume that Tweety can fly. If we later discover that Tweety is an ostrich, we may retract this assumption. Hence inferences based on such assumptions may need to be modified or rejected on the basis of later revelations. One way of accounting for default reasoning is through non-monotonic logic (see Non-monotonic Reasoning[182]), which allows conclusions to be rejected when more information is added to the premises. In non-monotonic logics, the usual set of axioms used in deriving theorems is augmented by a set of default rules which serve to infer conclusions that cannot otherwise be derived.

Formally, a default logic is a pair (W, D), where W is a set of first-order sentences and D is a set of default rules, each of which has the form:

$$\frac{a : b}{c}$$

which can be read as 'from a and the inability to prove $\neg$b, infer c'. The first order sentences a, b, and c are called the prerequisite, the justification, and the consequent of the default, respectively. The defaults can be viewed as extending the knowledge (as contained in W) that we have about the world. Any acceptable set of beliefs about that world is called an 'extension'. Default theories may have none, exactly one, or many extensions.

Reiter, R., *A logic for default reasoning*, Artificial Intelligence **13** (1980), 81–132, also in Readings in Non-monotonic Reasoning (Ginsberg, M.L., ed.), Morgan Kaufmann, Los Altos, CA, 1987, pp. 68–93.

Helen Lowe

65 Definite Clause Grammar

(DCG)

A DCG consists of a set of rules in a notation based on Logic Programming[153]. Each rule is similar to a Context-free Rule[54], with variables to show connections between the constituents involved, and an arbitrary test or action can be appended to the rule (for example, to handle number-agreement). DCGs are an extension of the grammar-rule notation in Prolog[216], and can be used to parse a string simply by interpreting them in a manner similar to the execution of a Prolog program.

Pereira, F. and Warren, D.H.D., *Definite clause grammars for language analysis—a survey of the formalism and a comparison with augmented transition networks*, Artificial Intelligence **13** (1980), 231–278.

Graeme Ritchie

66 Delayed Evaluation

A technique for generating a piece of program as a sequence of instructions (or as a composition of functions) from a set of clauses which specify constraints on data items. The data items are represented by tokens, some of which are replaced by list structures representing 'recipes' or promises to construct the given items from other items. The clauses are run through an interpreter using Data-directed Control[61] which finds an execution path. Instead of evaluating the computed result directly, it builds up a composed function or recipe. This composed function can be used in various ways:

- it can be transformed and optimised,
- if a target representation for the data is supplied, it can be compiled into code which freezes a fast execution path (or database search) that has been found by the interpreter,
- it can be interpreted directly using Lazy Evaluation[143] in a Function Programming Language[101].

There is a connection with Least Commitment[211] techniques, in that these, too, delay solution of a particular goal until sufficient constraints are available.

Gray, P.M.D., *Logic, Algebra, and Databases*, Ellis Horwood, Chichester, 1984.

Gray, P.M.D. and Moffat, D.S., *Manipulating descriptions of programs for database access*, Proceedings of IJCAI-83, 1983, pp. 21–24.

Todd, S.J.P., *The Peterlee relational test vehicle*, IBM Systems Journal **15** (1976), 304.

P.M.D. Gray

67 Demon

(Antecedent Theorem, If-added Method)

A part of a program which is automatically triggered when particular condition(s) occur; for instance a knowledge manipulation program might implement inference rules as demons. Whenever a new piece of knowledge is input, some of the demons (depending on the nature of the particular piece of knowledge) would activate and create additional pieces of knowledge by applying their sets of inference rules to it. These new pieces of knowledge in turn might result in more demons begin activated, as the inference process filters down through chains of logic. In the meantime, the main program can continue with whatever is its primary task. See Data-directed Control[61].

Nilsson, N.J., *Principles of Artificial Intelligence*, Tioga Pub. Co., Palo Alto, CA, 1980, and Springer-Verlag, Berlin, 1982.

AIWORD.DOC file (online: ARPA-net)

68 Dempster-Shafer Theory

(*Belief Functions*)

A theory of evidence potentially suitable for knowledge-based systems, especially when domain knowledge can be stated in categorical terms and the impact of each item of evidence described as an assignment of probabilities to a set of propositions. They include strict taxonomic hierarchies, terminological definitions and descriptions of deterministic systems (e.g., electronic circuits). The probability $m(A)$ attached to A is the degree to which the evidence supports A, but $m(A)$ need not be $1 - m(-A)$, since evidence can lend support to a hypothesis while having no direct bearing on its negation. These 'basic probabilities' can be visualised as probability masses constrained within the subset (of worlds) with which they are associated, but free to move over every point there. From these basic probabilities we can derive upper and lower probabilities (Dempster) or belief functions and plausibilities (Shafer). The belief measure $Bel(A)$ is the probability that A logically follows from the evidence, given that all items of evidence are non-contradictory. This notion of 'belief' often behaves differently than $P(A)$, the probability that A is true. Basic probabilities are combined using Dempster's Rule, which is valid for independent items of evidence.

Shafer, G., *A Mathematical Theory of Evidence*, Princeton University Press, Princeton, NJ, 1976.

Pearl, J., *Probabilistic Reasoning in Intelligent Systems: Networks of Plausible Inference*, Morgan Kaufmann, San Mateo, CA, 1988.

Judea Pearl

69 Dependency Directed Backtracking

(*Selective Backtracking*)

An alternative to Chronological Backtracking[71] where the backtrack point (the choice point that control is passed back to on failure) is determined by the nature of the failure. That is, the choice that caused the failure is undone whereas in chronological backtracking it is simply the last choice that is reconsidered. Some of the work done since the faulty choice may be independent of that choice, and with appropriate techniques much of this work can be retained.

Stallman, R.M. and Sussman, G.J., *Forward reasoning and dependency-directed backtracking in a system for computer-aided circuit analysis*, Artificial Intelligence **9** (1977), 135–196.

Pereira, L.M. and Porto, A., *Selective backtracking*, Logic Programming (Clark, K.L. and Tärnlund, S.A., eds.), Academic Press, London, 1982, also in APIC Studies in Data Processing, no. 16, pp. 107–117.

Lincoln Wallen

70 Depth-first Parsing

Amounts to a Depth-first Search[71] of the tree of alternatives which arise in the parsing process. Clearest in the context of Top-down[279] parsing of a Context-free Grammar[54]. Suppose our grammar includes the following rules:

$$X \rightarrow Q, R, S.$$
$$X \rightarrow U, V, W.$$

Suppose that in the course of analysing a string we are trying to find an X, which leads us to look for a Q, which we find. A choice then confronts us—do we look now for an R following the Q, or do we rather look for a U at the same place we found the Q? Given the possibility of ambiguity, either or both course may succeed. The first choice, which leads to a backtracking parser, is depth-first. The second choice, which leads to a pseudo-parallel parser, is Breadth-first[28]. If the candidate string was in fact ambiguous, in the depth-first case we would find Q R S X_1 U V W X_2 in that order, whereas in the breadth-first case we would find Q U R V S W X_1 X_2.

Winograd, T., *Language as a Cognitive Process, Volume 1: Syntax*, Addison-Wesley, Reading, MA, 1983.

Pereira, F.C.N. and Shieber, S.M., *Prolog and Natural Language Analysis*, CSLI Lecture Notes no. 10, University of Chicago Press, Chicago, IL, 1987, p. 32.

Gazdar, G. and Mellish, C., *Natural Language Processing in Prolog: An Introduction to Computational Linguistics*, Addison-Wesley, Wokingham, 1989.

Henry Thompson

71 Depth-first Search

(*Chronological Backtracking*)

An uninformed graph searching strategy which searches the graph by exploring each possible path through it until either the required solution or a previously encountered node is encountered. The nodes are expanded in order of depth: with the deepest node expanded first and nodes of equal depth expanded in an arbitrary order. To prevent searching of an infinite path, a depth-bound is usually fixed and nodes below this depth are never generated, thus the strategy is neither guaranteed to produce the shortest path to the solution if one exists, nor to find a solution even if one exists. (See also Iterative Deepening[138].)

Nilsson, N.J., *Principles of Artificial Intelligence*, Tioga Pub. Co., Palo Alto, CA, 1980, and Springer-Verlag, Berlin, 1982.

Korf, R.E., *Search: a survey of recent results*, Exploring Artificial Intelligence (Survey Talks from the National Conferences on Artificial Intelligence) (Shrobe, H.E., ed.), Morgan Kaufmann, San Mateo, CA, 1988, pp. 197–237 (Chapter 6).

Dave Plummer

72 Dependency Grammar

(*DG, Dependency Syntax*)

Dependency grammar (DG) describes the syntactic structure of sentences in natural languages in terms of links (dependencies) between individual words rather than constituency trees. The concept of dependency apparently originated with the Arabs and was adopted into Latin traditional grammar in the Middle Ages. The key claim of DG is that every phrase has a most prominent element (the head) which determines its syntactic properties. Modifiers and complements of the head are called dependents. The fundamental relation in dependency grammar is between head and dependent. One word (usually the main verb) is the head of the whole sentence; every other word depends on some head, and may itself be the head of any number of dependents. The rules of grammar then specify what heads can take what dependents (for example, adjectives depend on nouns, not on verbs). Practical DGs distinguish various types of dependents (complement, adjunct, determiner, etc.). DG, unlike Phrase Structure Grammar[54], does not take the constituency of syntactic elements (how are words grouped into phrases and how these phrases are grouped into bigger phrases, etc.) to be a central notion but rather concentrates on the relations between ultimate syntactic units. DG is therefore concerned with meaningful links, i.e., semantics. The syntactic representations postulated by the grammar are dependency trees. The nodes in a dependency tree correspond to word forms and the links between nodes represent binary relations between them. Thus, in dependency trees there are no non-terminal categories. The linear order between words is not stated in dependency trees and all meaningful distinctions expressed by word order variation should be explicitly stated using appropriate nodes and labelled arcs connecting them. DG has proved particularly useful in treating syntactic phenomena in free word-order languages (like Latin, the Slavic languages, etc.). It has found use in Machine Translation systems because dependency structures are seen to be very close to semantic structures.

Tesnière, L., *Elements de Syntaxe Structurale*, Klincksieck, Paris, 1959.

Melćuk, I., *Dependency Syntax: Theory and Practice*, State University of New York Press, Albany, NY, 1987.

Nicolas Nicolov

73 Deterministic Parsing

Non-determinism arises in a parsing process which proceeds on a word by word or constituent by constituent basis because of (often local) ambiguity in the grammar and lexicon. The degree of non-determinism can therefore be reduced by expanding the focus of attention to include more than just the 'next' constituent or word. Marcus has recently generated considerable interest with the claim that a small expansion of this sort allows English to be parsed deterministically in all but a few cases and in those it is claimed that people have difficulty as well.

Marcus and others pursuing this approach employ a stack and buffer parsing technique, in which words (and sometimes completed constituents) are accessed through a fixed length (typically 3 or 5 items) buffer, and partially completed constituents are held in a stack. Grammar rules are represented by condition-action pairs, where the conditions refer to the buffer and the stack, and the actions effect changes therein. Rules are grouped together for the purpose of activation and deactivation—only some groups of rules are active at any given time.

Strict determinism is ensured by the finiteness of the buffer and by requiring that all structures constructed at any point in the analysis must figure in the final result. It is an open question 'how much' of English, or other natural languages, can be analysed with grammars which can be parsed in this fashion.

Marcus, M., *A Theory of Syntactic Recognition for Natural Language*, MIT Press, Cambridge, MA, 1980.

Henry Thompson

74 Difference of Gaussians

This function is composed of the difference of two Gaussian distributions and approximates $\nabla^2 G$ (the Laplacian[142] of a Gaussian), the operator which Marr and Hildreth proposed to be optimal for Edge Detection[85] in images. It also describes the 'mexican hat' weighting function of the receptive fields of retinal ganglion and LGN cells.

Marr, D. and Hildreth, E., *Theory of edge detection*, Proceedings of the Royal Society, London **B207** (1980), 187–217, also in Computer Vision: Principles (Kasturi, R.J. and Jain, R.C., eds.), IEEE Computer Society Press, Los Alamitos, CA, 1991, pp. 77–107.

T.P. Pridmore, S.R. Pollard, S.T. Stenton

75 Discrimination Learning

If a production system contains overly general rules, they will lead to errors of commission (i.e., the rule will fire when it should not). The discrimination learning method responds to such errors by comparing the rejection context (the situation in which the mistake was made) to a selection context (the last situation in which the rule applied correctly). Based on the differences it finds between the two contexts, the method creates one or more variants of the overly general rule that contain additional conditions or otherwise restrict its generality. The new rules may still be overly general, in which case the technique is applied recursively until variants are found that match only in the desired contexts. The discrimination method can learn disjunctive concepts, and when combined with a strengthening process to direct search through the rule space, it can deal with noise and learn heuristically useful rules despite incomplete representations.

Anderson, J.R. and Kline, P.J., *A learning system and its psychological implications*, Proceedings of IJCAI-79, vol. 1, 1979, pp. 16–21.
Langley, P., *Language acquisition through error recovery*, Cognition and Brain Theory (1982).

Pat Langley

76 Discrimination Net

(*Discrimination Tree*)

A mechanism for allocating an input data item to its class by applying successive tests for different individual predicates; the terminal nodes of the net represent the results to be returned for the various possible sequences of predicates. A discrimination net (D-net) is thus a nest of IF...THEN...ELSE tests all applicable to one data item, or more formally, a binary, directed, acyclic graph with unary predicates at non-terminal nodes. An example of the use of a discrimination net of this basic kind is in natural language generation in choosing an output word for an input meaning representation. The basic mechanism can be extended by, for instance, using n-ary rather than binary graphs, with the corresponding replacement of simple feature tests by more complex branch selection functions, by the use of variables in the data item descriptions and net patterns, and by the use of sophisticated means of indexing. With such extensions, a net can be used, for example, to implement a PLANNER-style database optimised for retrieving individual assertions.

Discrimination nets have an obvious attraction when the set of classes involved is high; but clearly a prerequisite for their effective application is being able to identify clear test sequences for data items.

Charniak, E., Riesbeck, C.K., McDermott, D.V. and Meechan, J.R., *Artificial Intelligence Programming*, Lawrence Erlbaum Associates, Hillsdale, NJ, 1988 (second edition).

Karen Sparck Jones

77 Discourse Representation Theory

(*DRT*)

Discourse Representation Theory (DTR) treats the semantics and logic of natural languages. DRT represents larger linguistic units like discourses or texts and describes how syntactic form determines linguistic meaning. It provides a dynamic conception of meaning based on the observation that a human recipient of a discourse can process discourse on-line in an incremental fashion and the fact that new pieces of discourse are interpreted against the context established by the already processed discourse. Interpretation in DRT (i.e., the identification of meaning) proceeds in two stages. First, the incoming information is integrated sentence by sentence (incrementally) into the Discourse Representation Structure (DRS) for a discourse. By this integration process extra information, e.g., referential relations, may be added which does not form part of the compositional semantics of the individual sentences. Thus, a new piece of discourse updates the representation of the already processed discourse and the meaning of a linguistic expression consists both in its update potential and its truth-conditional import in the resulting representation. A completed DRS, i.e., a DRS which will not be further expanded, can be interpreted as a formula in Predicate Logic[208]. Intuitively a DRS can be conceived as a partial model representing the information conveyed by a discourse. DRT provides a better treatment of anaphora resolution and quantifiers than Montague Grammar[170].

Kamp, H. and Reyle, U., *From Discourse to Logic*, Kluwer Academic Publishers, London, 1993.

Nicolas Nicolov

78 Distributed Problem Solving

(*Distributed Planning*)

When a problem can be divided into independent subproblems (cf. Problem Reduction[267]), concurrent solution of them is possible and may be advantageous. For example, Kornfeld's ETHER language permits experimentation with concurrency in Heuristic Search[116], and Smith has implemented a contract net system motivated by the metaphor of manager-contractor linkage. The task of generating a plan may similarly be distributed over several concurrent planners, provided that each knows enough about the activities of the others to ensure consistency.

Kornfeld, W.A., *ETHER—a parallel problem solving system*, Proceedings of IJCAI-79, vol. 1, 1979, pp. 490–492.

Davis, R. and Smith, R.G., *Negotiation as a metaphor for distributed problem solving*, Artificial Intelligence **20** (1983), 63–109.

Jim Doran

79 Distributional Analysis

A statistical method in natural language processing, which involves examining the contexts in which certain strings tend to appear, in order to derive heuristics to classify substrings.

Garside R., Leech G. and Sampson, G., *The Computational Analysis of English: A Corpus-Based Approach*, Longman, London, 1987.

John Beavan

80 DSSS*

DSSS* is the Dual of the SSS* algorithm[259]. Whereas SSS* searches through the space of partial solution (or personal) trees, DSSS* searches through the space of partial adversary trees. An adversary tree can be formed from an arbitrary Game Tree[10] by pruning the number of branches at each MIN node to one. Such a tree represents a complete strategy for MIN, since it specifies exactly one MIN action for every possible sequence of moves that might be made by MAX.

Like SSS*, DSSS* never examines a node that Alpha/beta Pruning[8] would prune, and may prune some branches that alpha/beta would not, but it suffers from the same increased memory and storage requirements.

Shinghal, R., *Formal Concepts in Artificial Intelligence*, Chapman and Hall, London, 1992, pp. 549–569.

Ian Frank

81 Dynamical Systems

A Dynamical System is a system with a number of properties which evolve in parallel over time. They can be described as differential equations, using geometrical descriptions or computational models. Examples of dynamical systems include cellular automata, neural networks, spreading activation networks, Hopfield Nets[123], Boltzman Machines[263] and Genetic Algorithms[104].

Steels, Luc, *Artificial intelligence and complex dynamics*, Concepts and Characteristics of Knowledge-based Systems: Selected and Reviewed Papers from the IFIP TC 10/WG 10.1 Workshop, North-Holland, Amsterdam, 1989, pp. 369–404.

Martin Westhead

82 Dynamic Programming

Dynamic programming is a Template Matching[273] technique which allows the template to be 'stretched' in a non-linear fashion to find a best fit with the input pattern. It is a general purpose technique with many applications. In speech research, it has been used for recognition of words from a limited vocabulary within spoken utterances (cf. Dynamic Time Warping[83]).

Ney, H., *The use of a one-stage dynamic programming algorithm for connected word recognition*, Readings in Speech Recognition (Waibel, A. and Lee, K.F., eds.), Morgan Kaufmann, San Mateo, CA, 1990, pp. 188–196.

Sakoe, H. and Chiba, S., *Dynamic programming algorithm optimization for spoken word recognition*, Readings in Speech Recognition (Waibel, A. and Lee, K.F., eds.), Morgan Kaufmann, San Mateo, CA, 1990, pp. 159–165.

Steve Isard

83 Dynamic Time Warping

A method of processing speech utterances. Two approaches appear in the literature, one that generates Linear Predictive Coding[149] and autocorrelation coefficients from the speech utterances and the other that, using Dynamic Programming[82], compares the test utterance with the reference utterances and finds the best match. The method used is constrained endpoint with 2-to-1 range of slope.

Rabiner, L.R., Rosenberg, A.E. and Levinson, S.E., *Considerations in dynamic time warping algorithms for discrete word recognition*, IEEE Trans. ASSP **26** (1978), no. 6, 575–582.

Andrej Ljolje

84 Earley's Algorithm

The first published algorithm to bring the worst-case asymptotic time complexity of recognition for unrestricted Context-free Grammars[54] down to cubic order in the number of words in the string to be parsed. Makes use of a well-formed substring table. See Caching[29].

Earley, J., *An efficient context-free parsing algorithm*, Communications of the ACM **13** (1970), no. 2, 94–102, also in Readings in Natural Language Processing (Grosz, B.J., Sparck Jones, K. and Webber, B.L., eds.) Morgan Kaufmann, Los Altos, CA, 1986, pp. 25–33.

Henry Thompson

85 Edge Detection

The purpose of edge detection is to locate places in images which correspond to informative scene events, such as a shadow or the obscuring boundary of an object. The projection of such events often produces steep gradients and discontinuities in image intensity, so the basic methods of edge detection are:

- differentiate the image and threshold the resulting gradient image,
- differentiate the image twice and find the zero-crossings in the second derivative, and
- Template Matching[273].

Computationally, this normally involves Convolution[59] of a neighbourhood operator with the image at every point in the image. Very many operators have been implemented, but because of the difficulty of precisely specifying the goal of edge detection, it is difficult to compare operator performance. The latest theories of edge detection involve applying operators to the same image at different resolutions in resolution cones in an attempt to provide a better description of the intensity changes present in the image. See also Boundary Detection[25].

Marr, D. and Hildreth, E., *Theory of edge detection*, Proceedings of the Royal Society, London **B207** (1980), 187–217, also in Computer Vision: Principles (Kasturi, R.J. and Jain, R.C., eds.), Los Alamitos, CA, IEEE Computer Society Press, 1991, pp. 77–107.

Bob Beattie

86 Epsilon Semantics

(*Probabilistic Entailment, Preferential Entailment*)

A Non-monotonic Inference[182] system in which defaults are interpreted as statements of high conditional probabilities, infinitesimally close to zero or one. The valid consequences are those statements that would acquire arbitrarily high probabilities whenever the probabilities assigned to the premises are made sufficiently high. Syntactical derivations require the solution of a series of propositional satisfiability problems. The system is suitable for maintaining consistency and enforcing specificity ordering, but does not support chaining and contraposition.

Pearl, J., *Probabilistic Reasoning in Intelligent Systems: Networks of Plausible Inference*, Morgan Kaufmann, San Mateo, CA, 1988 (Chapter 10).

Pearl, J., *Probabilistic semantics for non-monotonic reasoning: a survey*, Proceedings of First Intl. Conf. on Principles of Knowledge Representation and Reasoning (KR'89), 1989, pp. 505–516.

Judea Pearl

87 Expert System Shell

A shell is a domain-independent Expert Systems 'framework', i.e., an inference engine with explanation facilities etc. but without any domain-specific knowledge. Provided the framework is a suitable one for a given problem, it may be possible by adding domain-specific knowledge to build a new Expert System with much less effort than beginning from scratch.

On the negative side, the use of a shell imposes a number of design decisions on, for example, the form in which knowledge is represented (it may be that only Production Rules[213] are catered for), and the means by which reasoning with uncertain knowledge is handled (e.g., Bayesian Inference[17], Dempster-Shafer Theory[68], etc.). These can considerably reduce the range of applicability of a particular shell.

Commercial shells are generally readily available for standard business computers (such as IBM PCs) but typically little or no attempt is made by suppliers to identify the range of applicability of their products.

Shells developed for research purposes (e.g., EMYCIN based on MYCIN, and KAS based on Prospector) generally offer more sophisticated facilities, but are frequently difficult to obtain outside the academic research community, are often poorly documented and even when available may be difficult to transfer to new systems.

Hayes-Roth, F., Waterman, D.A. and Lenat, D.B. (eds.), *Building Expert Systems*, vol. 1, Addison-Wesley, Reading, MA, 1983.

Max Bramer

88 Explanation-Based Learning

Explanation-Based Learning (EBL) techniques are knowledge-based and require a great deal of knowledge about the domain of the concept to be learned. Instead of providing several instances of a concept (as in Focussing[92]), only a single example is required for a new concept to be learned. The idea is that a strong domain theory not only helps in describing or *explaining* the new concept, but also guides the generalisation process, such that a great deal can be learned from a single example.

Various EBL techniques have been developed: constraint back propagation, explanation-based generalisation, Precondition Analysis[206]; mostly in domains where there is a strong body of domain knowledge.

DeJong, G. and Mooney, R., *Explanation-based learning: an alternative view*, Machine Learning **1** (1986), no. 2, 145–176.

Mitchell, T.M., Keller, R.M. and Kedar-Cabelli, S.T., *Explanation-based generalisation: a unifying view*, Machine Learning **1** (1986), no. 1, 47–80.

Roberto Desimone

89 Extended Gaussian Image

A method of object shape representation in which an object is represented as a function on the unit sphere, by mapping points on the object surface to the point on the unit sphere with corresponding normal direction. The value of the function on the unit sphere is the sum (integral for a planar surface) of the absolute values of the inverse of the Gaussian curvatures of the corresponding surface points. The integral of the function over the unit sphere is the object's surface area. The representation rotates in correspondence with the object and for convex objects is unique (disregarding translation).

Horn, B.K.P., *Robot Vision*, MIT Press, Cambridge, MA, 1986 (16).

R.M. Cameron-Jones

90 Fast Pattern Recognition Techniques

A technique for classifying very large bit vectors either by means of special purpose hardware or software. There are two phases: a 'learning' phase and a 'use' phase. During the 'learning' phase examples of bit vectors together with a classification are presented. During the 'use' phase unknown vectors are presented to the system and classified. The technique is based on the Bledsoe and Browning n-tuple methodology.

Aleksander, I. and Stonham, T.J., *A guide to pattern recognition using random-access memories*, IEEE Journal of Computers & Digital Techniques **2** (1979), no. 1.

Igor Aleksander

91 Feature Structures

(*FSs*)

Feature Structures are structured objects which consist of a set of attributes (features) and their corresponding values. FSs are much like frames in AI systems, records of imperative programming languages like C or Pascal, and the feature descriptions in standard theories of phonology, and more recently in syntax. The value of an attribute in a FS can in turn be a structured object (FS). FSs can be seen as labelled tuples. Two features (possibly at different levels of nesting) can share their value (can be token identical). This is called reentrancy or coreference. Because FSs allow reentrancy in general they are not labelled trees but Directed Acyclic Graphs (DAGs). The main operation on FSs is feature structure unification which takes two FSs and attempts to find the most general FS that is compatible with them. Unification of FSs is a simple form of merging. FSs can be seen as partial descriptions of (linguistic) objects—unification can always add new features (with their values) or further specify an embedded FS (see Graph Unification[113]).

Typed feature structures are like ordinary feature structures except that each feature structure (node in the DAG) has an associated type (or sort) that identifies what type of object the structure is modelling. The set of all types is partially ordered and each type has an associated list of valid features that can be specified for it. FSs have found wide use in theoretical linguistics: sorts in constraint-based grammars (such as HPSG[114]), f-structures (in Lexical Functional Grammar (LFG), feature bundles, feature matrices or categories (in GPSG[107]), functional structures (in FUG[100]), DAGs (in PATR-II). Recent computational linguistics frameworks are based on feature structures (ALE, TFS, CUF, TDL, etc.). Even programming languages use FSs as their basic data structures (ψ-terms in LIFE are a generalisation of typed feature structures).

Gazdar G. and Mellish, C.S., *Natural Language Processing in PROLOG*, Addison-Wesley, Wokingham, 1989 (Section 7.2; pp.230-235; Section 7.10).

Nicolas Nicolov

92 Focussing

(*Version Space*)

A technique utilised by concept learning programs to record knowledge about the concept that is being learnt and to facilitate the updating of this knowledge as more information becomes available to the program. For this method the description space must consist of a number of attributes that a given example might have. These are conceptually represented either as sets of descriptions or as lattices, each branch of which represents a possible value for the attribute. Using the later representation, the technique records information by placing marks in these lattices to indicate which possible attribute values are not included in the concept being learnt. Upper marks indicate that all parts of the tree above the mark are definitely outside the concept (i.e., this is the maximally general description) and a lower mark indicates that all parts of the tree below the mark are definitely inside the concept (i.e., this is the maximally specific description). Between these marks there could be a grey region which is the area that the program is uncertain about. As the program gains more information the marks are moved to update the knowledge: positive instances of a concept are used to make the maximally specific description more general by moving the lower marker up, whilst negative instances are used to make the maximally general description more specific by moving the upper mark down. Once the upper and lower marks on all of the description trees coincide, the concept is completely specified and the program can report completion. The program can also detect failure if the marks become crossed (i.e., a lower mark above an upper one).

This technique can handle most conjunctive concepts well but encounters difficulty handling disjunction, these difficulties may be surmountable by using multiple copies of the description space.

Young, R.M., Plotkin, G.D. and Linz, R.F., *Analysis of an extended concept-learning task*, Proceedings of IJCAI-77, vol. 1, 1977, p. 285.

Mitchell, T.M., *Generalization as search*, Artificial Intelligence **18** (1982), 203–226.

Bundy, A. and Silver, B., *A critical survey of rule learning programs*, Proceedings of ECAI-82, 1982, pp. 151–157.

Dave Plummer, Maarten van Someren

93 Formant

A speech signal can be described through formants; these are the principal resonances of the vocal tract, that determine which overtones of the fundamental larynx frequency will be allowed to pass through into the speech signal, and which will be damped. Each position of the articulators (the lips, tongue, jaw, etc.) gives rise to a characteristic pattern of formants. A speech signal can therefore be represented in terms of formant patterns which change over time, both for purposes of recognition and synthesis-by-rule.

Markel, J.D. and Gray Jr., A.H., *Linear Prediction of Speech*, Springer-Verlag, New York, 1976.

Steve Isard

94 Formant Synthesis

Basic sounds of a language can be characterised by their Formant[93] frequencies. The formant pattern for a synthetic utterance can be created by interpolating between the formant values of the sounds composing the desired utterance. A speech wave can be computed from such a formant pattern, to be played through a digital-to-analogue converter, or the formant pattern can be used directly to excite a set of resonators, acting as an electrical analogue of the vocal tract.

Klatt, D.H., *Software for a cascade/parallel formant synthesizer*, Journal of Acoustic Society of America **67** (1980), 971–995.

Steve Isard

95 FORTRAN

FORTRAN is the programming language considered by many to be the natural successor to Lisp[150] and Prolog[216] for AI research. Its advantages include:

- it is very efficient for numerical computation (many AI programs rely heavily on number-crunching techniques).
- AI problems tend to be very poorly structured, meaning that control needs to move frequently from one part of a program to another. FORTRAN provides a special mechanism for achieving this, the so-called GOTO statement.
- FORTRAN provides a very efficient data structure, the array, which is particularly useful if, for example, one wishes to process a collection of English sentences each of which has the same length.

Coan, J.S., *Basic FORTRAN*, Hayden Book Co, Indianapolis, 1980.

Aloysius Hacker

96 Forward Search

(*Data-driven Search, Bottom-up Search, Forward Chaining*)

A State Space[260] can be searched from the initial state to the goal state, applying Operators[187] forward (forward search); alternatively the state space can be searched from the goal to the initial state (Backwards Search[16]).

Barr, A. and Feigenbaum, E.A. (eds.), *The Handbook of Artificial Intelligence*, vol. 1, William Kaufmann, Los Altos, CA, 1981.

Maarten van Someren

97 Fourier Transform

A method that transforms a (spatial or temporal) signal into the frequency domain by exploiting the fact that a signal can be represented as the sum (or integral) of a (usually infinite) series of weighted sinusoidal functions.

The Fourier transform is useful for analysing and understanding the influence of linear operations on temporal or spatial signals. A good qualitative understanding about how a certain operation influences the signal in the frequency domain often helps to estimate limitations or possibilities that are otherwise not so easily recognised. This is most important for low-level vision applications. Fourier transforms are the most important tool for linear systems theory.

The Fourier transform is quite tedious computationally: i.e., the computational complexity of most one-dimensional transforms is $O(Nlog(N))$ and two-dimensional transforms is $O(N^2log(N))$ where typically $N > 250$. Hence, it is usually impractical to implement the transform in software, although multiprocessor systems are now being used. Alternatives include use of special-purpose hardware or optical techniques.

Bracewell, R.N., *The Fourier Transform and its Applications*, McGraw-Hill, New York, 1986 (second edition, revised).

Gaskill, J.D., *Linear Systems, Fourier Transforms, and Optics*, Wiley, New York, 1978.

Fritz Seytter

98 Frame

(*Schema*)

A principle for the large-scale organisation of knowledge introduced by Minsky, originally in connection with vision, but more generally applicable. A simple example is the GUS air trip frame, a structure with slots for the various elements of a trip, e.g., passenger, source, destination, date, etc., which are instantiated in any particular application of the frame. Frames may be arbitrarily complex, and have procedures attached to the slots. Default values for the slots are helpful when frames are used in the absence of full instantiation data. The character of frames suggests a hierarchical organisation of sets of frames, but non-hierarchical filling of one frame slot by another is possible. Frame structures are often not deemed to imply temporal or causal relations between their slots, and are thus contrasted with Scripts[239] but community usage in this respect is very inconsistent: one person's frame is another person's script, and vice versa. The main problem with systems with multiple frames is frame selection and replacement.

Minsky, M., *A framework for representing knowledge*, The Psychology of Computer Vision (Winston, P.H., ed.), McGraw-Hill, New York, 1975, pp. 211–277.

Karen Sparck Jones

99 Functional Data Model

An entity-relationship data model with relationships expressed as functions from arguments to a single entity value or to a set of entities. Derived functions allow the definition of arbitrary new relationships in terms of existing ones. Different views of data can be presented by defining appropriate derived functions. Updating of derived relationships is supported through procedures explicitly provided by the user. Data manipulation languages, such as DAPLEX, have been added to the Functional Data Model to provide the notion of looping through entity sets to perform operations. The Functional data model incorporates many of the ideas in earlier Semantic Net[244] systems, the if-needed and if-added theorems of the PLANNER and CONNIVER languages, and a range of data models.

Shipman, D.W., *The functional data model and the data language DAPLEX*, Readings in Artificial Intelligence and Databases (Mylopoulos, J. and Brodie, M.L., eds.), Morgan Kaufmann, San Mateo, CA, 1989, pp. 168–184.

Austin Tate

100 Functional Unification Grammar

(*FUG*)

Functional grammar is a declarative expression of the relation between surface structure and functional structure in a natural language equally suitable for use in parsing and generation. The functional structure (or Feature Structure[91]) is a directed acyclic graph of features and their values indicating the functional roles played by sentence constituents, e.g., subject, modifier ... Functional grammar has been adopted by a number of language generation systems.

Kay, M., *Parsing in functional unification grammar*, Natural Language Parsing: Psychological, Computational and Theoretical Perspectives (Dowty, D.R., Karttunen L., Zwicky, A.M., eds.), Cambridge University Press, Cambridge, 1985, pp. 251–278.

McKeown, K.R., Paris, C.L., *Functional unification grammar revisited*, Proceedings of the 25th Annual Meeting of the Association for Computational Linguistics (1987), 97–103.

Henry Thompson

101 Functional Programming Language

(*Applicative Language, Purely Functional Language*)

A language where the value of an expression (its meaning) is determined solely by the values of its constituent parts. Such languages have no assignment statements, and make use of higher-order functions to avoid the need for imperative control constructs. Functional programs are often easier to reason about than their imperative counterparts, and subexpressions may be evaluated in any order, including in parallel, a property which can be exploited by concurrent architectures.

Henderson, P., *Functional Programming—Application and Implementation*, Prentice-Hall, Englewood Cliffs, NJ, 1980.

Kevin Mitchell

102 Fuzzy Logic

Fuzzy logics were created to deal with vague sentences and partial degrees of truth. The truth-value of a sentence in a fuzzy logic is a value in the interval $[0,1]$, rather than *true* or *false*. Intuitively, assigning a truth-value 0 to a sentence corresponds to the concept of the sentence being *false*, assigning a truth-value 1 corresponds to the sentence being *true*, and intermediate values (i.e., values between 0 and 1) represent *distances* between the truth-value of the sentence and the values *true* and *false*.

Fuzzy truth-values are propagated by means of *triangular norms and conorms* and *symmetric inversion functions*. For example, if the sentence a has truth-value μ_a the sentence b has truth-value μ_b, then the sentence $a \wedge b$ can be defined to have the truth-value $\mu_\wedge = min\{\mu_a, \mu_b\}$, the sentence $a \vee b$ can be defined to have the truth-value $\mu_\vee = max\{\mu_a, \mu_b\}$ and the sentence $\neg a$ can be defined to have the truth-value $\mu_\neg = 1 - \mu_a$.

Dubois, D. and Prade, H., *An introduction to possibilistic and fuzzy logics*, Non-Standard Logics for Automated Reasoning (Smets, P., Mamdani, E.H., Dubois, D. and Prade, H., eds.), Academic Press, London, 1988, pp. 287–326.

Flávio Corrêa da Silva

103 Fuzzy Set Theory

An extension of conventional set theory in which the grade of membership for an element in a set taking a value anywhere in the range [0,1], instead of 0 or 1 only. Developed as a means of avoiding the complexity in descriptions of subjective or ill-understood processes.

Zadeh, L.A., *Outline of a new approach to the analysis of complex systems and decision processes*, IEEE Trans. of Systems, Man and Cybernetics **3** (1973), 28–44.

Mamdani, E.H. and Gaines, B.R., *Fuzzy Reasoning and its Applications*, Academic Press, London, 1981.

Janet Efstathiou

104 Genetic Algorithm

(*Evolutionary Algorithm*)

A Genetic Algorithm (GA) is a population-based optimisation strategy which finds use as a robust and often very effective technique for fast convergence to near-optima or optima of ill-understood fitness landscapes. Key concepts are the notion of a 'chromosome' and its 'fitness'. A chromosome is a (usually) fixed-length string of symbols which can be interpreted as a potential solution to the problem in hand; its fitness is a measure of how close this chromosome is to a perfect solution. A GA works via continual update of a collection (called a 'population') of chromosomes, which are initially randomly generated. Updating comprises three steps: (i) One or more chromosomes are selected from the population; (ii) One of a set of genetic operators is applied to the selected chromosome(s), yielding one or more 'offspring'; (iii) The offspring enter the population, replacing less fit members. The process iterates until some stopping criterion is met (e.g., a good enough or ideal solution is found), or all chromosomes in the population are the same. Variations on the GA theme range across a vast space of possible ways of doing steps (i), (ii), and (iii). Step (ii), in which genetic operators are applied, is particularly important in terms of how the search is conducted. The commonest operators are crossover, in which segments of each of two 'parent' chromosomes are interchanged to produce one or more offspring, and mutation, in which a small part of a single parent is altered. Real-world applications of GAs usually benefit from the design of operators incorporating domain specific knowledge.

Lawrence Davis, *The Handbook of Genetic Algorithms*, Van Nostrand Reinhold, New York, 1991.

Dave Corne

105 Game Theoretic Semantics

The game-theoretic interpretation of logic is due to Hintikka. In this approach, any attempt to establish the truth or falsity of an expression, S, in an interpreted language, L, is correlated with a two-person, zero-sum, perfect information game, G(S), played according to the rules of L. These games can be thought of as constituting 'idealised processes of verification'. Informally, one can think of the two players as oneself and 'Nature', and the game consists of one seeking support for S, while Nature looks for a refutation.

Hintikka, J., *Logic, Language Games and Information: Kantian Themes in the Philosophy of Logic*, Clarendon Press, Oxford, 1973.

Patric Chan

106 Generalised Cylinders

(*Generalised Cone*)

A currently popular method for three-dimensional shape description. Generalised cylinders are a class of objects obtained by extending the definition of a cylinder. An ordinary cylinder is the volume swept out by a circular disc moving perpendicular to a straight line segment that passes through its centre called the axis or spine of the cylinder. A generalised cylinder is then obtained by extending the definition to include things like:

- A curved spine.
- The radius of the disc varying as a function of the position along the spine.
- The cross section being some planar figure other than a circle.
- The cross section held at a non-perpendicular angle to the spine.

Cohen, P.R. and Feigenbaum, E.A. (eds.), *The Handbook of Artificial Intelligence*, vol. 3, Pitman, London, 1982.
Marr, D., *Analysis of occluding contour*, Proceedings of the Royal Society, London **B197** (1977), 441–475.

Luis Jenkins

107 Generalised Phrase Structure Grammar

(*GPSG*)

The GPSG was a move away from the more traditional computational syntactic framework of Transformational Grammar. The main theoretical point of this formalism is that it is equivalent to Context-free Grammars[54]. The formalism differs from conventional phrase structure grammars in that it:

- Describes syntactic categories in terms of lists of feature and value pairs, e.g., `[N +, V -, AGR [PER 3, PLU -]]` to represent a third person singular noun.
- Instead of conventional mother and daughters type rule, two equivalent notations are used: Immediate Dominance (ID) specifying which categories can dominate which daughters; and Linear Precedence (LP) rules specifying the order of these daughters.
- Feature passing conventions, specify default 'copying' of features up and down the parse tree.
- Meta-rules[162] can be used to define new ID/LP rules from other existing ones thus the meta-rules, and basic ID/LP rules effectively specify, if compiled out, a set of context free grammar rules.

Following GPSG a number of other similar syntactic theories were born under the global title of Unification Grammars[287], e.g., Lexical Functional Grammar (LFG), Head-Driven Phrase Structure Grammar (HPSG)[114] and Categorial Unification Grammar[35] (CUG).

Gazdar, G., Klein, E., Pullum, F. and Sag, I., *Generalised Phrase Structure Grammar*, Blackwell, Oxford, 1985.

Alan W. Black

108 Generate and Test

(*Model Directed Search*)

Problem solving can be viewed as the process of generating solutions from observed or given data. Unfortunately, it is not always possible to use direct methods (i.e., to go from data to solution directly). Instead, we often have to use indirect or model-based methods. If we have a model of the real world, we can test the results we would expect from hypothetical solutions against those that we desire or observe in the real world. The generate and test search technique is to generate some plausible solutions from some initial conditions and test them using the model. If necessary, we return/backtrack to the generate stage, forming more solutions in the light of our test results then test these.

This is repeated until an acceptable solution is found. Examples of systems using generate and test search are Meta-Dendral and Casnet.

Feigenbaum, E.A., Buchanan, B.G. and Lederberg, J., *On generality and problem solving: a case study using the Dendral program*, Machine Intelligence 6 (Meltzer, B. and Michie, D., eds.), Edinburgh University Press, Edinburgh, 1971, pp. 165–190.

Weiss, S.M., Kulikowski, C.A., Amarel, S. and Safir, A., *A model based method for computer-aided medical decision making*, Artificial Intelligence **11** (1978), 145–172.

Robert Corlett

109 Goal Structure

(*Holding Periods, Plan Kernels, Plan Rationale, Teleology*)

Information about a plan which represents the purpose of the various plan parts in terms of the goals or subgoals they are intended to achieve. This can be contrasted with the Plan Structure[202] which contains details which may obscure information important to the finding of causes for planning or execution failures.

The STRIPS Macro Operator[155] Tables[188]) used such information for plan execution. Goal structure has been used in a linear planner (INTERPLAN) and a non-linear planner (NONLIN) to guide search. It can be viewed as a Meta-planning[160] technique.

Drummond, M. and Tate, A., *AI planning*, Knowledge Engineering, Volume 1: Fundamentals (Adeli, H., ed.), McGraw-Hill, New York, pp. 157–199.

Austin Tate

110 Golay Neighbourhoods

A set of local patterns, devised by Golay for transformations on hexagonally connected binary images. They fall into the general class of morphological transforms ('hit or miss'). Each pattern is a hexagon of 'true', 'false' and 'do not care' entries. When a pattern matches the neighbourhood of a pixel in the image that pixel can be marked (to detect special points like line ends), set true (to 'thicken' the image) or set false (to 'thin' or 'skeletonise' the image). See Local Grey-level Operations[151].

Generalisations of such operations are now used for processing images with square connectivity.

Golay, M.J.E., *Hexagonal parallel pattern transformations*, IEEE Transactions on Computers **C-18** (1969), no. 8.

Dave Reynolds

111 Gradient Space

A two-dimensional space used to represent surface orientation in terms of its vertical and horizontal components. Thus the plane surface $-Z = P.X + Q.Y + C$ is represented by the point (P, Q) in gradient space. The steepness of the surface is $\sqrt{P^2 + Q^2}$ and the direction of the slope is $\tan^{-1}(Q/P)$. Such a representation does not make explicit the spatial location or extent of surface planes. Convex and concave edges and curvatures will be represented by lines in the gradient space, in the case of planar surface discontinuity edges these will be perpendicular to the edge in the image space, with order along the line determined by the convexity/concavity.

Draper, S.W., *The use of gradient space and dual space in line drawing interpretation*, Artificial Intelligence **17** (1981), 461–508.

T.P. Pridmore, S.R. Pollard, S.T. Stenton

112 Graph Matching

Relational structures are often represented as graphs with nodes connected by labelled arcs. The problem of matching two such structures is known as graph matching and can be achieved in a number of ways. Firstly a one-to-one mapping can be achieved between the nodes and arcs of each of the two graphs. This is known as graph isomorphism. Alternatively a one-to-one mapping can be achieved between one graph and a *subgraph* of the other. This is known as double subgraph isomorphism. Sometimes a perfect mapping between graphs is not required, some arcs and nodes being considered unimportant and omitted from the matching process.

Harary, F., *Graph Theory*, Addison-Wesley, Reading, MA, 1969.

H.W. Hughes

113 Graph Unification

Many recent grammar theories, such as Generalised Phrase Structure Grammar[107], Lexical Functional Grammar, and Categorial Grammar[35] may be expressed using the notion of Directed Acyclic Graphs (DAGs), Feature Structures[91], or Attribute-Value Matrices (AVMs). The grammatical information associated with an expression can be thought of as a set of partial functions (or *features*) defined on it, whose values may be either atomic, or a further set of partial functions. Alternatively, this may be represented by a DAG where the root node represents the expression, the directed edges are labelled by the name of a feature, and the nodes where they end represent the corresponding value.

For instance, the following feature structure may represent an expression with syntactic category NP, and with an *agreement* feature, *number*, whose value is singular:

$$\begin{bmatrix} cat : np \\ agr : [num : sing] \end{bmatrix}$$

Information is then passed between linguistic entities by the operation of *Graph Unification*, which informally consists of 'combining' two graphs to produce a third one containing the information of the first two, and nothing else, provided this information is not contradictory (in which case unification fails).

For instance, the feature structure described above may unify with the similar feature structure below (which has a 3rd person agreement feature):

$$\begin{bmatrix} cat : np \\ agr : [person : 3] \end{bmatrix}$$

The result of the unification will then be the following feature structure:

$$\begin{bmatrix} cat : np \\ agr : \begin{bmatrix} num : sing \\ person : 3 \end{bmatrix} \end{bmatrix}$$

Shieber, S.M., *An Introduction to Unification-Based Approaches to Grammar*, CSLI Lecture Notes no. 4, University of Chicago Press, Chicago, IL, 1986.

John Beavan

114 Head-driven Phrase Structure Grammar

(*HPSG*)

Head-Driven Phrase Structure Grammar (HPSG) is an integrated theory of natural language syntax and semantics. HPSG is a lexicalist theory—structure is determined chiefly by the interaction between highly articulated lexical entries and parameterised universal principles of grammatical well-formedness. The phrase structure rules are reduced to a small number of highly general and universally available phrase structure (or immediate dominance) schemata.

The linguistic objects in HPSG (i.e., signs) are modelled using typed Feature Structures[91]. The theory posits the existence of a single stratum of linguistic representation with different levels of linguistic structure expressed as values of appropriate attributes. HPSG is non-transformational and non-derivational—attributes in the linguistic structure are related using the notion of *structure sharing*, that is, token identity between substructures of a given structure, in accordance with lexical specifications or grammatical principles (or complex interactions between the two). HPSG doesn't make use of tree configurational notions (like c-command). The grammar is viewed as a declarative system of constraints which determine what the well-formed expressions of the language are. Because of the nature of typed feature structures, and more specifically, because of the monotonicity of the Unification[113] operation, the order in which linguistic constraints are applied is irrelevant. Thus, HPSG is a competence model of language. HPSG is a very widespread grammatical framework among computational linguists. Implementations of HPSG grammars are naturally embedded within constraint-based, unification frameworks (see Unification Grammars[287]).

Pollard, C. and Sag, I., *Head-Driven Phrase Structure Grammar*, University of Chicago Press, Chicago, IL, 1994.

Nicolas Nicolov

115 Heterarchy

In a hierarchical process organisation, data passes through a sequence of analytical or interpretive modules each of which acts independently of the rest. In a heterarchical system, modules may invoke other modules in the series to help with, for example, disambiguation of data. Winograd's program SHRDLU, in which syntactic analysis could make use of semantic modules on knowledge about the world was the most famous example. Shirai and others designed heterarchic image understanding programs. Heterarchy fell into disrepute when the work of Horn, Barrow and Tenenbaum suggested that far more disambiguation can be done autonomously by low levels than was previously thought. The fashion will change again when it is realised that in poor viewing conditions more sophisticated process organisation is required.

Clowes, M.B., *Man the creative machine*, The Limits of Human Nature (Benthall, J., ed.), Allen Lane, London, 1973.

Aaron Sloman

116 Heuristic Search

(*Best-First Search*)

A technique for State Space[260] searching, with the state space represented as a graph. It uses domain-specific knowledge expressed as a numerical evaluation function which assigns a number to each node of the graph. At each stage of the search, heuristic search develops the tip node with the best numeric score. Tip nodes may be stored on an agenda in order of numeric score. See A* Algorithm[2], Alpha/Beta Pruning[8], B* Algorithm[15], Branch-and-bound Algorithms[26], and Minimax[164] for examples of heuristic search.

Nilsson, N.J., *Principles of Artificial Intelligence*, Tioga Pub. Co., Palo Alto, CA, 1980, and Springer-Verlag, Berlin, 1982.

Korf, R.E., *Search: a survey of recent results*, Exploring Artificial Intelligence (Survey Talks from the National Conferences on Artificial Intelligence) (Shrobe, H.E., ed.), Morgan Kaufmann, San Mateo, CA, 1988, pp. 197–237 (Chapter 6).

Dave Plummer

117 Hidden Markov Models

Hidden Markov modelling is a powerful stochastic approach to speech recognition. It is based on the assumption that the speech signal may be approximated by a first-order Markov process. Although this assumption is erroneous, their tractability, firm grounding in statistics and the existence of a powerful and provably convergent training algorithm have made hidden Markov models the dominant technique in speech recognition. Hidden Markov models are 'hidden' since each state in the model does not necessarily correspond to a particular portion of the speech signal, but contains an output probability distribution over the input space.

Rabiner, L.R. and Juang, B.H., *An introduction to hidden Markov models*, IEEE ASSP Magazine **3** (1986), no. 1, 4–16.

Steve Renals

118 Hierarchical Models

At the highest level a hierarchical model consists of a number of complex parts positioned within some global coordinate reference frame. Descending the hierarchy, each part is decomposed into more simple parts each with its own local coordinate frame. At each level the coordinate frame transformation between part and subpart is specified. By allowing degrees of freedom in the transformation it is possible to represent flexible joints between parts. Finally, at the lowest level of the hierarchy, the parts are decomposed into primitives (the basic building blocks) usually either volumes or surfaces (e.g., Generalised Cylinders[106]). An example of this would be the modelling of a human hand. At the highest level would be the complete model of the hand. This would decompose into the palm, fingers and thumb. A finger would decompose further into the primitives representing them.

Brooks, R.A., *Symbolic reasoning among 3-D models and 2-D images*, Artificial Intelligence **17** (1981), no. 1, 285–349.

H.W. Hughes

119 Hierarchical Planning

(*Plan Refinement, Top-down Plan Elaboration*)

The technique by which a hierarchical plan structure is produced by successive generation, from top to bottom, of its component levels. One variant of the technique developed by Sacerdoti makes use of criticalities assigned to operator preconditions. More generally, it was found by Tate, for example, that domain specific plan elaboration rules are required. Top-down plan elaboration is closely related to automatic top-down programming techniques (see Barstow).

Sacerdoti, E.D., *Planning in a hierarchy of abstraction spaces*, Artificial Intelligence **5** (1974), 115–135, also in Readings in Planning (Allen, J., Hendler, J. and Tate, A., eds.) Morgan Kaufmann, San Mateo, CA, 1990, pp. 98–108.

Tate, A., *Generating project networks*, Proceedings of IJCAI-77, vol. 2, 1977, pp. 888–893, also in Readings in Planning (Allen, J., Hendler, J. and Tate, A., eds.) Morgan Kaufmann, San Mateo, CA, 1990, pp. 291–296.

Barstow, D.R., *An experiment in knowledge-based automatic programming*, Artificial Intelligence **12** (1979), no. 2, 73–119, also in Readings in Artificial Intelligence (Rich, C. and Waters, R.C., eds.) Morgan Kaufmann, San Mateo, CA, 1986, pp. 135–156.

Jim Doran

120 Hierarchical Synthesis

A technique used in the computer-based recognition of structure or objects. In a hierarchical synthesis recognition system, primitive or simpler features are recognised first, and then joined together to recognise larger objects in a bottom-up manner. For example, one might recognise the handle and bowl of a coffee cup first, and then verify that these were in positions compatible with being part of the same cup. This form of matching is often used as part of a Model-based Vision[167] system.

Fisher, R.B., *From Surfaces to Objects: Computer Vision and Three Dimensional Scene Analysis*, John Wiley and Sons, Chichester, 1989.

Robert B. Fisher

121 High-emphasis Filtering

A method for sharpening images. The differentiation of an image is grossly interpreted in the frequency domain as filtering that emphasizes higher frequency components. As any linear operator in the spatial domain can be converted into an equivalent transfer function in the frequency domain, a linear operator designed to emphasize abrupt changes in intensity can be implemented by a transfer function designed to emphasize areas of high frequency. See also Fourier Transform[97].

Cohen, P.R. and Feigenbaum, E.A. (eds.), *The Handbook of Artificial Intelligence*, vol. 3, William Kaufmann, Los Altos, CA, 1982, pp. 206–215.

Luis Jenkins

122 Hill Climbing

Hill climbing is a search method for finding a maximum (or minimum) of an evaluation function. It considers the local neighbourhood of a node, evaluating each neighbour node, and next examines those nodes with the largest (or smallest) values. Unlike other search strategies that use evaluation functions (e.g., the A* algorithm[2] or informed Depth-first Search[71]) hill climbing is an irrevocable scheme: it does not permit us to shift attention back to previously suspended alternatives, even though they may have offered a better alternative than the ones at hand. This property is at the heart of both its computational simplicity and its shortcomings; it requires very little memory since alternatives do not need to be retained for future consideration. However, it is not guaranteed to lead to a solution, since it can get stuck on a plateau or a local optimum, or even wander on infinite uncontrolled paths, unless the guiding evaluation function is very informative.

Pearl, J., *Heuristics: Intelligent Search Strategies For Computer Problem Solving*, Addison-Wesley, Reading, MA, 1984.

Rina Dechter

123 Hopfield Nets

The Hopfield Network, introduced in the early 1980s by John Hopfield, can be used (i) as an Associative Memory[12], (ii) to perform Constraint Satisfaction[52], or (iii) as a model of a physical (spin glass) system. Hopfield not only formulated a network model, but also contributed a set of analysis tools which employ the concept of an energy function.

Hopfield's network is implemented as a fully connected array of nodes which contain symmetrical connections (or weights). Each node calculates the weighted sum of its binary inputs (minus a threshold value) and applies a step-function to the result in order to determine its output state. When employed as an associative memory, a recipe is used to calculate the weights as a function of the input patterns to be stored (i.e., inputs are associated with each other). Recall is performed by presenting the net with an unknown pattern (e.g., a degraded version of a stored pattern) as a set of starting states for all the nodes. The network then proceeds to cycle through a succession of states until it converges on a stable solution. The operation of the network can also be understood from an energy perspective. This involves creating an energy function such that the programmed patterns are basins of attraction within some energy landscape which describes the possible states of the system. In order to perform combinatorial and optimisation problems (e.g., the Travelling Salesman Problem), the network cannot be programmed using the desired attractors (because they comprise the solution), but rather is programmed directly from the energy function of which a minimum is sought.

Hopfield, J.J., *Neural networks and physical systems with emergent collective computational abilities*, Proceedings of the National Academy of Sciences (USA) **81** (1984), 3088–3092.

Hertz, J., Krough, A. and Palmer, R.G., *Introduction to the Theory of Neural Computation*, Addison-Wesley, Redwood City, CA, 1991.

Ashley Walker

124 Hough Technique

The Hough technique is a method of detecting parametric image features, such as curves. A point on a curve undergoes a transform to form a path in parameter space that represents all possible curves that could pass through that point. The parameter space is represented by an array whose elements are incremented each time a path passes across them. By repeating the transform for each point, local peaks are formed in parameter space representing points lying on the same curve in image space. The height of the peak indicates the number of points on the curve. To detect straight lines, for a point at position (x, y) the transform $r = x\cos\theta + y\sin\theta$ is usually used to produce a sinusoidal curve in (r, θ) parameter space. This technique can be used on any curve that can be parameterised, though the amount of computation required increases rapidly with the number of parameters.

Ballard, D.H. and Brown, C.M., *Computer Vision*, Prentice-Hall, Englewood Cliffs, NJ, 1982, pp. 123–131.

H.W. Hughes

125 Image Data Structures

A wide variety of data structures are used to represent images. At the low level, raw grey-level images or binary images are represented by arrays of pixels (with square, triangular or hexagonal connectivity). Object boundaries are described by Fourier descriptors or strings (Freeman chain code, symbolic strings). The adjacency of object regions is described by graph structures such as the region adjacency graph. Finally, hierarchical or Pyramidal[221] data structures which describe an image at a series of different levels or resolutions have proved useful (e.g., Quad Trees[222]).

Tanimoto, S. and Klinger, A., *Structured Computer Vision: Machine Perception Through Hierarchical Computation Structures*, Academic Press, New York, 1980.

Dave Reynolds

126 Image Flow Estimation

The estimation of the two-dimensional velocity field which maps the image intensity pattern at one time instant on to that at the succeeding time instant. The image flow is equivalent to the Optical Flow[190] if the image intensity value for the (changing) image point corresponding to a point in the world remains constant. Methods for recovering image flow are usually based on the motion constraint equation (which relates the local image velocity to local spatial and temporal intensity derivatives) and thus are hindered by the aperture problem—the local irrecoverability of velocity along a direction of zero intensity gradient. For this reason methods for flow recovery at points other than intensity extrema require the imposition of further constraints, and Regularisation[231] methods are commonly used for flow estimation on curves or over regions.

Nagel, H.H., *On the estimation of optical flow: relations between different approaches and some new results*, Artificial Intelligence **33** (1987), 299–324.

R.M. Cameron-Jones

127 Image Segmentation

Segmentation attempts to identify important features in an image, such as lines or regions of a uniform colour. In low-level segmentation, pixels are grouped together using techniques like thresholding or (for edges) Convolution[59]. In high-level segmentation, expectations from models are used to guide the use of such low-level feature detectors. See also Model-based Vision[167].

Fu, K.S. and Mui, J.K., *A survey on image segmentation*, Pattern Recognition **13** (1981), 3–16.

Brady, M., *Computational approaches to image understanding*, ACM Computer Survey **14** (1982), no. 1, 3–71, also in Machine Vision: The Advent of Intelligent Robots (Brady, M., ed.), Addison-Wesley, Wokingham, 1986, pp. 7–66.

Dave Reynolds

128 Incidence Calculus

(*Probabilistic Logic*)

A multi-valued logic in which the value of a formula is a set of situations in which the formula is true. These situations can be possible worlds, models or members of any set of objects. The values of composite formulae are defined to be set theoretic combinations of the values of their sub-formulae, e.g., the value of a conjunction is the intersection of the values of the conjuncts. The formulae can be assigned a number, e.g., a probability, by applying an arithmetic function to their assigned sets, e.g., the sum of the probabilities of each member of the set. In this way incidence calculus can be used to represent the uncertainty of a formulae, e.g., as a probability. The main advantage of this indirect route of assigning probabilities is that the correlation between two formulae is implicitly represented by the degree of intersection between formulae. Thus it is not necessary to assume conditional independence. When a new conclusion is deduced from some hypotheses the value of the conclusion is not completely determined by the values of the hypotheses, but it is possible to calculate tight upper and lower bounds on value of the conclusion.

Bundy, A., *Incidence calculus: a mechanism for probabilistic reasoning*, Journal of Automated Reasoning **1** (1985), 263–284.

Alan Bundy

129 Inductive Logic Programming

Inductive Logic Programming (ILP) encompasses a set of techniques from machine learning and Logic Programming[153]. ILP systems develop predicate descriptions from examples and background knowledge. The examples, background knowledge and final descriptions are all described as logic programs. A unifying theory of Inductive Logic Programming is being built up around lattice-based concepts such as refinement, Least General Generalisation[145], Inverse Resolution[135] and most specific corrections. In addition to a well established tradition of learning-in-the-limit results, some results within Valiant's PAC-learning framework have been demonstrated for ILP systems. Presently successful applications areas for ILP systems include the learning of structure-activity rules for drug design, finite-element mesh analysis design rules, primary-secondary prediction of protein structure and fault diagnosis rules for satellites.

Muggleton, S.H., *Inductive logic programming*, New Generation Computing **8** (1991), 295–318.

Muggleton, S.H. (ed.), *Inductive Logic Programming*, Academic Press, London, 1992.

Steve Muggleton

130 Inferno

Inferno is a conservative approach to uncertainty. It is based on a probabilistic approach to uncertainty, but, unlike most similar schemes, does not make any assumptions about relationships between propositions (e.g., independence). It is, however, possible to assert such relationships in Inferno if they exist. The method is essentially based on upper and lower probabilities (see Dempster-Shafer Theory[68]) which give upper and lower bounds on the probability of each proposition. Each time one of these bounds changes, the bounds on related propositions are checked and suitably modified to satisfy a set of propagation constraints. These constraints are based on the inequality:

$$max(P(A), P(B)) \leq P(A \text{ or } B) \leq P(A) + P(B)\,.$$

One unique feature of Inferno is its ability to detect and suggest corrections for contradictions and inconsistencies in its data. If the propagation constraints cannot provide valid values for the bounds, Inferno invokes a method similar to Dependency Directed Backtracking[69] to suggest changes to the data that would remove the inconsistency.

Quinlan, J.R., *Inferno: a cautious approach to uncertain inference*, Expert Systems: Techniques, Tools, and Applications (Klahr, P. and Waterman, D.A., eds.), Addison-Wesley, Reading, MA, 1986, pp. 350–390.

Robert Corlett

131 Interactions Between Sub-solutions

Whenever a planner makes the assumption that conjunctive goals can be solved independently (either one after the other or in parallel) there is likely to be interference between the partial solutions. Techniques have been developed to recognise and correct for these interactions between solutions to conjunctive goals.

Sussman's HACKER solved problems by assuming an ordered solution was feasible. It then had critics to recognise interactions and HACKER was often able to repair the plan by re-arranging the steps in the plan. In his INTERPLAN program, Tate's approach was to abstract the original goals and to determine holding periods over which these could be assumed to be true. INTERPLAN analysed this Goal Structure[109] with a view towards ordering the approach taken to solve the sub-goals to ease conflict situations. Waldinger developed an approach called 'goal regression' in which a solution to one goal was built and then the plan was constructively modified to achieve the further plans (by moving new goals backwards through a partial plan to a position where they did not interfere). A scheme similar to that used by Waldinger was incorporated in WARPLAN, a planner written in Prolog[216].

All the above-mentioned planners produce their plans as a linear sequence of actions. The Non-linear Planner[181] NONLIN, also dealt with interactions between sub-goals in plans which are produced as partially ordered networks of actions.

Drummond, M. and Tate, A., *AI planning*, Knowledge Engineering, Volume 1: Fundamentals (Adeli, H., ed.), McGraw-Hill, New York, pp. 157–199.

Tate, A., *Interacting goals and their use*, Proceedings of IJCAI-75, 1975, pp. 215–218.

Austin Tate (adapted from Mark Stefik)

132 Interlingua

One of the two mainstream approaches to machine translation which involves using an intermediate language as a pivot between the languages to be translated, and provides at least a semantic representation of the text to be translated. The intermediate representation is usually some logic-based notation for the semantics.

In an interlingual approach to machine translation the main components would be a mapping from the source language to the interlingua (analysis component), and another from the interlingua to the target language (generation component). The interlingua approach is mostly used in systems translating between many languages, where its main advantage is that only one analysis and one generation component have to be built for each language involved. Compare with Transfer[280].

Tucker, A.B., *Current strategies in machine translation research and development*, Machine Translation: Theoretical and Methodological Issues (Nirenburg, S., ed.), Cambridge University Press, Cambridge, 1987.

John Beavan

133 Interval Calculus

Interval calculus stands for a description of temporal relationships in which time is treated in terms of intervals. There are 13 possible primitive relations between pairs of intervals, e.g., 'X before Y', 'X meets Y'. The relationships between intervals are maintained in a network where the nodes represent individual intervals and the arcs label possible relations between the intervals. This allows one to construct non-primitive relations. There are rules of transitivity, allowing a computationally effective inference process, e.g., 'if (X during Y) and (Y meets Z) then (X before Z)'.

Allen, J., *Toward a general model of action and time*, Artificial Intelligence **23** (1984), 123–154.

K. Sundermeyer

134 Intrinsic Images

The term was first used by Barrow and Tenenbaum (1983) to refer to a registered 'stack' of retinotopic maps each of which makes explicit the value of a certain property 'intrinsic' to the surfaces in the scene. The intrinsic images all have the same viewer centred coordinate system but carry information about different surface properties, such as: surface orientation, depth, reflectance, colour, texture and Optical Flow[190] (this last intrinsic image describes the instantaneous velocity flow field in the scene). The computation of the intrinsic images is non-trivial, many of them are underdetermined when considered independently but global consistency constraints can be cooperatively exploited, e.g., surface boundary information carried by the reflectance image can be used to constrain the computation of the surface orientation from shading information.

Over the last five years there have been considerable advances in understanding of the problems of computing the different intrinsic images. Their particular importance is that they are a vital stage in the computation of a representation intermediate between the lowest levels of image processing, whose descriptions are essentially 2-D pictorial or iconic descriptions of the scene, and the higher levels of processing which describe the shapes of objects in terms of an viewer independent Object-centred Coordinate[184] system. The intrinsic images are the first representation at which information concerning the 3-D structure of the scene is made explicit. Object recognition schemes that have attempted to recover 3-D shape descriptions directly from 2-D shape descriptions have, to put it mildly, struggled. See also $2\frac{1}{2}$-D Sketch[1].

Brady, M., *Computational approaches to image understanding*, Machine Vision: The Advent of Intelligent Robotics (Brady, M., ed.), Addison-Wesley, Wokingham, 1986, pp. 7–66.

Barrow, H.G. and Tenenbaum, J.M., *Computational vision*, Proceedings of IJCAI-83, 1983, pp. 39–74.

Jon Mayhew

135 Inverse Resolution

This machine learning technique involves the construction of First-order Logic[208] clauses by the inversion of single Resolution[234] steps. Two inverse resolution operators exist, the V and the W. The V operator takes the consequent clause and one of the antecedent clauses of a resolution step and constructs the other antecedent. The W operator takes the consequents of two resolutions which have a common antecedent and constructs all three antecedents, thus introducing a new predicate symbol. Inverse resolution is provably capable of constructing a wide class of theories, including Recursive Theories[225], given a set of ground unit examples.

Muggleton, S. and Buntine, W., *Machine invention of first-order predicates by inverting resolution*, Proceedings of the Fifth International Conference on Machine Learning, Morgan Kaufmann, San Mateo, CA, 1988, pp. 339–352.

Steve Muggleton

136 IS-A Hierarchy

(*Inheritance Hierarchy, Type Hierarchy*)

Strictly a straightforward manifestation of class membership, found useful in knowledge representation because property specifications for classes need only be explicitly indicated once, i.e., at the highest level, since they are automatically inherited by subclasses. IS-A relationships are often found in a Semantic Net[244]. This form of IS-A relationship must be clearly distinguished from IS-A relationships between individuals and the classes of which they are members. This second version of the IS-A relationship is also found in semantic nets. The characteristic of semantic nets whereby properties of nodes connected by transitive links are inherited by adjacent nodes is called property inheritance, and so the IS-A link is a property inheritance link.

Fahlman, S.E., *NETL, A System for Representing and Using Real-world Knowledge*, MIT Press, Cambridge, MA, 1977.

Karen Sparck Jones

137 Island Parsing

The adaptation of Augmented Transition Network[13] parsing to deal with the problems presented by, for example, speech where the terminal symbols of the grammar cannot be certainly identified in correct linear sequence in the input. Bidirectional parsing is initiated from any plausibly identified node in the network, an island, providing hypotheses to assist the identification of uncertain input items. However parsing is complicated by the fact that tests and actions may be context dependent, and may not be executable because the required information for leftward input items is not yet available.

Bates, M., *The theory and practice of augmented transition network grammars*, Natural Language Communication with Computers (Lecture Notes in Computer Science, vol. 63) (Bolc, L., ed.), Springer-Verlag, Berlin, 1978, pp. 191–260.

Karen Sparck Jones

138 Iterative Deepening

An uninformed graph search algorithm which is a good compromise between the efficiency of Depth-first Search[71] and the admissibility of Breadth-first Search[28]. Iterative deepening performs a complete search of the Search Space[260] (often using a depth-first search strategy) up to a maximum depth d. If no solutions can be found up to depth d, the maximum search depth is increased to $d+1$, and the search space is traversed again (starting from the top node). This strategy ensures that iterative deepening, like breadth-first search, always terminates in an optimal path from the start node to a goal node whenever such a path exists (this is called admissibility); but it also allows implementation as an efficient depth-first search. Iterative deepening is optimal in both time and space complexity among all uninformed admissible search strategies.

At first sight it might look as if iterative deepening is inefficient, since after increasing the cut-off depth from d to $d+1$, it redoes all the work up to level d in order to investigate nodes at level $d+1$. However, since typical search spaces grow exponentially with d (because of a constant branching factor), the cost of searching up to depth $d+1$ is entirely dominated by the search at the deepest level $d+1$: If b is the average branching rate of the search space, there are $b^{(d+1)}$ nodes at depth $d+1$, which is the same as the total number of nodes up to depth d. In fact, it can be shown that among all uninformed admissible search strategies, iterative deepening has the lowest asymptotic complexity in both time ($O(b^d)$) and space ($O(d)$). Breadth-first search on the other hand is only asymptotically optimal in time, and is very bad (exponential) in space. The actual time complexity of breadth-first search (as opposed to the asymptotic complexity) is of course lower than that for iterative deepening (namely by the small constant factor $b/b-1$), but this is easily offset by the difference in space-complexity in favour of iterative-deepening. Thus, iterative-deepening is asymptotically optimal in both time and space, whereas breadth-first is asymptotically optimal only in time and really bad in space, while the actual complexities of iterative-deepening and breadth-first are very close.

Iterative deepening can also be applied to informed search strategies, such as A*[2]. This modified version of A* is again optimal in both time and space among all informed admissible search strategies.

Korf, R.E., *Depth-first iterative-deepening: an optimal admissible tree search*, Artificial Intelligence **27** (1985), no. 1, 97–109.

Korf, R.E., *Search: a survey of recent results*, Exploring Artificial Intelligence (Survey Talks from the National Conferences on Artificial Intelligence) (Shrobe, H.E., ed.), Morgan Kaufmann, San Mateo, CA, 1988 p, pp. 197–237 (Chapter 6).

Frank van Harmelen

139 Junction Dictionary

Within the blocks world domain, only a restricted set of edge types of vertices, and thus line junctions in the image, exist. A junction dictionary can be constructed for line labellings of each member of this set. Of all the possible line labellings (which would imply a potentially large dictionary), only a small subset are in fact physically possible. See also Relaxation Labelling[233] and Line Labelling[148].

Waltz, D., *Understanding line drawings of scenes with shadows*, The Psychology of Computer Vision (Winston, P.H., ed.), McGraw-Hill, New York, 1975, pp. 19–91 (Chapter 2).

T.P. Pridmore, S.R. Pollard, S.T. Stenton

140 Kinematics

Analysis of the kinematics of mechanisms can provide a technique for rational design of manipulators and workspaces. Mechanical arrangements for robot manipulators vary widely among operational robots, the most common configurations being best described in terms of their coordinate features: Cartesian, spherical, and articulated. In a Cartesian robot, a wrist is mounted on a rigid framework to permit linear movement along three orthogonal axes, rather like a gantry crane or (for two axes) a graph plotter; the resulting workspace is box-shaped. The cylindrical robot has a horizontal arm mounted on a vertical column which is fixed to a rotating base. The arm moves in and out; a carriage moves the arm up and down along the column, and these two components rotate as a single element on the base; the workspace is a portion of a cylinder. The spherical robot is similar to the turret of a tank: the arm moves in and out, pivots vertically, and rotates horizontally about the base; the workspace is a portion of a sphere. An articulated robot is more anthropomorphic: an upper arm and forearm move in a vertical plane above a rotating trunk. The limbs are connected by revolute joints; the workspace approximates a portion of a sphere. For all robots, additional degrees of freedom are provided at the extremity of the arm, at the wrist. Wrists generally allow rotation in two or three orthogonal planes. To make proper use of a robot arm, transformations between encoded axis values (joint angles, etc.) and more convenient coordinate systems must be computed at high speed. Transforming a set of axis values to a position and orientation in space is called the forward kinematics transformation. The reverse transformation is used to convert a desired position and orientation in space into commanded axis values.

Engleberger, J.F., *Robotics in Practice: Management and Applications of Industrial Robots*, Kogan Page, London, 1980.

Paul, R.P., *Robot Manipulators: Mathematics, Programming and Control*, MIT Press, Cambridge, MA, 1981.

Groover, M.P., Weiss, M., Nagel, R.N. and Odrey, N.G., *Industrial Robotics: Technology, Programming, and Applications*, McGraw-Hill, New York, 1986.

W.F. Clocksin

141 Lambda Calculus

Various formal systems based on that invented by Church to formalise the properties of functions acting on arguments and being combined to form other functions. This involves 'lambda-abstraction'. The function f given by:

$$f(x) = x + 1,$$

can be written using lambda-abstraction as:

$$f \;=\; \lambda x.x + 1,$$

so that:

$$f(1) = (\lambda x.x + 1)(1) = 2.$$

Application is written as juxtaposition, e.g., fx for $f(x)$. Terms made up using application and lambda abstraction can be manipulated in various ways, e.g., rename bound variables (alpha conversion), and rewrite $(\lambda x.fx)a$ as fa (beta reduction). Lambda calculus is the formalism that underlies Lisp[150].

Hindley, J.R. and Seldin, J.P., *Introduction to Combinators and the Lambda Calculus*, Cambridge University Press, Cambridge, 1986.

Barendregt, H.P., *The Lambda Calculus, its Syntax and Semantics*, North-Holland, Amsterdam, 1984 (revised edition).

Alan Smaill

142 Laplacian

The Laplacian is the lowest (second) order circularly symmetric differential operator. When convolved with a 2-D function (e.g., an image) it computes the non-directional second derivative of that function. See also Convolution[59].

Marr, D., *Vision: A Computational Investigation into the Human Representation and Processing of Visual Information*, W.H. Freeman, San Francisco, CA, 1982.

T.P. Pridmore, S.R. Pollard, S.T. Stenton

143 Lazy Evaluation

(*Call-by-need*)

A parameter passing mechanism in which the evaluation of an argument is postponed until its value is actually required. Used as an alternative to call-by-value where arguments to a function are evaluated before the function is called. The technique is particularly useful for the construction and manipulation of infinite data structures. See also Delayed Evaluation[66].

Peyton Jones, S.L., *The Implementation of Functional Programming Languages*, Prentice-Hall, Englewood Cliffs, NJ, 1987.

Kevin Mitchell

144 Learning from Solution Paths

Starting with only the legal conditions on a set of Operators[187], a strategy learning system can employ weak methods to search a problem space. Once a solution path has been found, it can be used to assign credit and blame to instances of these operators. If a move leads from a state on the solution path to another state on the solution path, it is labelled as a positive instance of the responsible operator. However, if a move leads from a state on the solution path to a state not on the path, it is marked as a negative instance. These classifications can then be input to a condition-finding mechanism (such as the generalisation, discrimination or Version Space[92] methods); which will determine the heuristically useful conditions under which each operator should be applied.

Sleeman, D., Langley, P. and Mitchell, T.M., *Learning from solution paths: an approach to the credit assignment problem*, AI Magazine **3** (1982), no. 2, 48–52.

Pat Langley

145 Least General Generalisation

Plotkin showed that there exists a dual of the most general unifier of two literals. Suppose that the most general unifying substitution (see Unification[286]) for two first-order literals, l_1 and l_2 is θ. Then $l_1\theta = l_2\theta$ is called the most general instance (mgi) of l_1 and l_2. Like the most general instance, the least general generalisation (lgg) of l_1 and l_2 does not always exist. However, whenever l_1 and l_2 have the same predicate symbol and sign their lgg is unique up to renaming of variables.

Generalisation over literals is defined by subsumption. l_1 subsumes l_2 whenever there exists a substitution θ such that $l_1\theta = l_2$. Plotkin showed that due to the uniqueness of the least general generalisation and most general instance, subsumption forms a infinite semi-lattice over first-order literals. Plotkin extends this result to show that the same holds for subsumption over clauses.

Subsumption can be defined relative to given first-order clausal background theory, B (see Clausal Form[41]). In this case, clause C relatively subsumes clause D whenever $B \wedge C \vdash D$. Plotkin studies the case in which $\vdash$ is taken to be Resolution-based[234] derivation in which each clause in $B \wedge C$ is used at most once. In this case there exists a unique relative least general generalisation (rlgg), C of two clauses D_1 and D_2. In Muggleton (1991) it is shown that the relative subsumption lattice is the same as the lattice searched by Inverse Resolution[135]. More specifically the most specific inverse resolvent to depth n is unique up to renaming of variables. The least general generalisation of the most-specific inverse resolvents of D_1 and D_2 is equivalent to the relative least general generalisation of D_1 and D_2.

Muggleton, S., *Inductive logic programming*, New Generation Computing **8** (1991), no. 4, 295–318.

Steve Muggleton

146 Lexical Access

(Dictionary Lookup)

There are various techniques for accessing lexical information ranging from the sensible, e.g., the use of a tree structure on a letter-per-branch basis, to the ridiculous, e.g., linear search. The larger the lexicon the more important it becomes to relate word frequency to the access method. In English, for example, a mere 100 word types account for over 50% of the word tokens in running text. It also becomes more important to take a general approach to the problems of inflected forms, e.g., past tenses, plurals, and spelling rules or Morphographemics[171].

Kay, M., *Morphological and syntactic analysis*, Linguistic Structures Processing (Zampolli, A., ed.), North-Holland, Amsterdam, 1977, pp. 131–234.

Kaplan, R.M. and Kay, M., *Word recognition* (1982), Xerox Palo Alto Research Center (Technical Report).

Henry Thompson

147 Library-based Planning

(*Variant Planning*)

Instead of the plan required for a given task being constructed de novo, a plan library is maintained and, given a particular task, a relevant plan is found in the library and adapted to the current need. This may involve storing, with the plans in the library, information about their derivation. The idea is similar to Case-based Reasoning[32].

Carbonell, J., *A computational model of analogical problem solving*, Proceedings of IJCAI-81, vol. 1, 1981, pp. 147–152.

Jim Doran

148 Line Labelling

Huffmann and Clowes conceptualised the task of interpreting straight line drawings in terms of attaching labels like 'concave', 'convex', 'occluding' to lines, interpreted as depicting object edges. See Junction Dictionary[139] and Relaxation Labelling[233].

Clowes, M.B., *On seeing things*, Artificial Intelligence **2** (1971), 79–116.

Aaron Sloman

149 Linear Predictive Coding

If one approximates the vocal tract as a series of fixed length tubes (which is equivalent to representing it as an all-pole digital filter) it becomes possible to predict successive samples of the speech wave as linear combinations of previous samples. The coefficients in the linear combination characterise the shape of the vocal tract. A sequence of sets of coefficients can be used to characterise the changing shape of the vocal tract over time. This representation is widely used because of the particularly efficient algorithms associated with it.

Linggard, R., *Electronic Synthesis of Speech*, Cambridge University Press, Cambridge, 1985 (Chapter 5).

Steve Isard

150 Lisp

Lisp (an acronym for *LIS*t *P*rocessing) is one of the primary AI programming languages. It was defined by John McCarthy in 1958. The reasons for its success are related not only to its early development but also to its technical characteristics:

- Applicative style: pure Lisp is a functional language whose underlying ideas are directly related to Lambda Calculus[141]. Thus, the increment function:

$$ADD1 = \lambda n.n + 1,$$

can be defined in Lisp as:

$$(DEFUN\ ADD1(n)(+\ n\ 1))$$

- Programs as data: programs are represented in the same data structure as other data, namely the list. A program can be manipulated as any other data structure.

The first characteristic is true only of what is called *pure Lisp*. All the existing different dialects of Lisp also have control structures for imperative programming. Among the most important dialects are: MACLISP (from MIT—see Winston and Horn), FRANZ LISP (a dialect of MACLISP developed at Berkeley—see Wilensky) and INTERLISP (developed by BBN and Xerox). Lately a big effort has been made to unify all the different dialects in an unique language called COMMON LISP (see Steele et al.).

Winston, P.H. and Horn, B.K.P., *LISP*, Addison-Wesley, Reading, MA, 1989 (third edition).

Wilensky, R., *LISPcraft*, W.W. Norton and Company, New York, 1984.

Steele, G.L., Fahlman, S.E., Gabriel, R.P., Moon, D.A., and Weinreb, D.L., *Common Lisp: The Language*, Digital Press, Burlington, 1984.

Fausto Giunchiglia

151 Local Grey-level Operations

(*Image Morphology*)

A class of transformations on grey-level image which replace each pixel (in parallel) by some function of its neighbouring pixels.

In pointwise operations the new pixel value is independent of the neighbourhood, for example simple grey scale re-mapping such as histogram equalisation.

Linear operations (Convolution[59]) are used for filtering (local averaging etc.) and feature detection (such as Edge Detection[85]).

Non-linear operations can be developed using local minimum and local maximum functions to replace summation. Such operations can also be regarded as grey level generalisations of binary morphological operations using the Fuzzy Logic[103] based replacement of AND by MIN and OR by MAX. Illustrative operations are 'shrink' and 'expand' (also called 'erode' and 'dilate'). In a shrink operation a 'true' pixel is changed to 'false' if there is a 'false' pixel in some defined neighbourhood of it. Similarly for expand. For example, if the neighbourhoods are defined as 3x3 squares then shrink will delete all edge pixels.

Some of these types of transformations have been generalised to three-dimensional binary images.

Serra, J., *Image Analysis and Mathematical Morphology*, Academic Press, London, 1982.

Dave Reynolds

152 Logic Grammar

The use of Logic Programming[153] for natural language processing. The approach views parsing as Theorem Proving[276] in a logic in which the lexicon and the grammar are interpreted as axioms and deduction rules. Prolog[216] can easily be used as a parser if the grammar rules are written as Horn clauses. However, it is generally acknowledged that First-order Logic[208] is not powerful enough for natural language applications, and many extensions have been recently proposed to overcome these deficiencies.

Dahl, V. and Saint-Dizier, P. (eds.), *Natural Language Understanding and Logic Programming, II*, Proceedings of the Second International Workshop on Natural Language Understanding and Logic Programming, Vancouver, Canada, 17-19 August, 1987, North-Holland, Amsterdam, 1988.

John Beavan

153 Logic Programming

The core idea of logic programming is that it is possible to provide a procedural semantics for statements in certain (usually classical) logical formalisms. In theory, programming is then reduced to providing a program specification that is represented as a set of axioms. This is handed over to a sound and complete Theorem Prover[276]. The theorem prover constructs a proof that there is a solution. If there is a solution then further useful information can usually be extracted from the proof. The main justification for such an approach is that classical logic is well understood and that, as a consequence, it is possible to provide systems which have a rigorous, theoretical foundation.

In order for such systems to be useful it is necessary that the formalism is sufficiently powerful and desirable that it is also sufficiently expressive. The best-known example of an attempt to realise such a system is Prolog[216] which is based on a Resolution[234] theorem prover. Programs are written in the Horn clause form of first-order Predicate Logic[208].

There is an extensive literature on the properties of Prolog (and some of its variations) vis-a-vis soundness and completeness. Note that the ideal has not been attained in practice as (i) the fixed control strategy built into most Prolog interpreters does not always find all solutions, (ii) some explicit control (procedural element) is usually required, and (iii) programs often require side effects which undermine the declarative semantics. There are also some difficulties in connection with extending the expressiveness of Prolog, in particular, the problems of Negation[178] and equality.

The term 'logic programming' is often taken to be synonymous with 'Prolog'. This is not correct, however, as the field of logic programming also includes the so-called committed choice languages. These languages use a Prolog-like formalism but are designed for various forms of parallel execution. Examples include PARLOG, GHC and CP. Such languages lack a clear declarative semantics. Other systems have been built which explore various extensions, alternative control regimes or methods for automating proof search other than resolution.

Balbin, I. and Lecot, K., *Logic Programming: A Classified Bibliography*, Wildgrass Books, Fitzroy, Australia, 1985.

Hogger, C.J., *Introduction to Logic Programming*, Academic Press, London, 1984.

Paul Brna

154 Logics of Knowledge and Belief

In order to construct a plan it is often necessary for an agent to take into account the possibility that certain pieces of knowledge may be required to execute the plan: knowledge which may be obtainable by performing certain actions. An agent may also need to take into account the knowledge and beliefs of other agents. So in designing an intelligent agent we need to employ some method for reasoning about knowledge and belief. A logic of knowledge or belief can be obtained by adding to some system of logic some means of expressing such facts as 'A knows that p' or 'A believes that p' where 'A' designates an agent and 'p' a proposition. So we could introduce names of agents and variables ranging over agents together with operators such as K and B, so that '$K(A,p)$' means that A knows that p and '$B(A,p)$' means that A believes that p.

In AI there are basically two approaches to the construction of a logic of knowledge or belief. There is a sentential approach where we associate with each agent a set of sentences which constitutes the agent's base beliefs. We say that an agent believes a proposition exactly if the agent can prove the proposition from these base beliefs. An agent knows a proposition if he or she believes the proposition and it is true. There is also an approach which construes the K operator as analogous to the L operator in Modal Logic[165]. Here we associate with each agent sets of possible worlds. We say that an agent believes a proposition in a given world just in case that proposition holds in every world accessible for an agent in a given world.

Moore, R.C., *A formal theory of knowledge and action*, Formal Theories of the Commonsense World (Hobbs, J.R. and Moore, R.C., eds.), Ablex Pub. Corp., Norwood, 1985, pp. 319–358.

Konolige, K., *A first order formulation of knowledge and action in a multi-agent planning system*, Machine Intelligence 10 (Hayes, J.E., Michie, D. and Pao, eds.), Ellis Horwood, Chichester, 1982, pp. 41–72.

Colin Phillips

155 Macro Operators

(*MACROPS, Triangle Table*)

It is possible to combine a sequence of Operators[187] to build a new one, a macro operator, that has the effect of the sequence. Its list of preconditions contains all preconditions of the first operator of the sequence, plus those of later operators in so far as they have not been satisfied by previous operators in the sequence. The ADD and DELETE LISTS are determined in the same way.

Macro operators can be represented like 'basic' operators and be added to the set of existing operators. This is a form of learning that will reduce search in new problems.

Fikes, R.E., Hart, P.E. and Nilsson, N.J., *Learning and executing generalized robot plans*, Artificial Intelligence **3** (1972), 251–288, also in Readings in Machine Learning (Shavlik, J.W. and Dietterich, T.G., eds.) Morgan Kaufmann, San Mateo, 1990, pp. 468–486.

Laird, J.E., Rosenbloom, P.S. and Newell, A., *Chunking in SOAR: the anatomy of a general learning mechanism*, Machine Learning **1** (1986), 11–46.

Maarten van Someren

156 Marker-passing

This technique involves performing parallel searches through a Semantic Net[244] by the placing of simple binary tags, called marks, on memory nodes. An example of the use of marker-passing would be to query a network as to the COLOR of an Elephant named CLYDE. The elements next to each of these would be marked, as would their neighbours in turn. The only node marked from each starting place would be GRAY, the correct answer. Much is now known about the limitations of such a technique and the mathematics of inheritance hierarchies. See also Symbolic Marker-passing[271].

Fahlman, S.E., *NETL: A System for Representing and Using Real World Knowledge*, MIT Press, Cambridge, MA, 1979.

James A. Hendler

157 Markov Fields

A probabilistic description of a system of variables that interact locally, organised as vertices in a graph. A Markov field encodes the assumption that each variable is influenced by only its nearest neighbours. In image processing, Markov fields are used to represent the correlations between adjacent picture cells. In Neural Networks[48], Markov fields represent the energy associated with each state.

Geman, S. and Geman, D., *Stochastic relaxation, Gibbs distributions and the Bayesian restoration of images*, IEEE Trans on PAMI **6** (1984), 721–741.

Judea Pearl

158 Maximum Cardinality Search

A method of ordering nodes in a graph which, whenever possible, guarantees the following condition: no two nodes will be linked to a successor node unless the two are linked together. The ordering consists of sequentially selecting a node that makes the largest number of connections with those selected earlier. This search yields close to optimal orderings for conducting adaptive consistency. It is also used in Constraint Satisfaction[52] and probabilistic reasoning problems to transform a network into a tree of clusters, thus facilitating local propagation and relaxation.

Dechter, R. and Pearl, J., *Tree clustering for constraint networks*, Artificial Intelligence **38** (1989), 353–366.

Tarjan, R.E. and Yannakakis, M., *Simple linear-time algorithms to test chordality of graphs, test acyclicity of hypergraphs and selectively reduce acyclic hypergraphs*, SIAM J. Computing **13** (1984), 566–579.

Rina Dechter

159 Means/ends Analysis

A technique for controlling search. Given a current state and a goal state, an Operator[187] is chosen which will reduce the difference between the two. This operator is applied to the current state to produce a new state, and the process is recursively applied to this new state and the goal state.

Newell, A. and Simon, H.A., *GPS, a program that simulates human thought*, Computers and Thought (Feigenbaum, E.A. and Feldman, J., eds.), McGraw-Hill, New York, 1963, pp. 279–293.

Alan Bundy

160 Meta-action Oriented Planning

(*Opportunistic Planning, Agenda Based Planning, Meta-planning*)

The operations that a planner performs upon the plans it is working with (e.g., expansion, reduction) are referred to as 'meta-actions'. A flexible and opportunity based choice of which meta-action to perform next can be achieved by maintaining a heuristic scheduler, where the tasks on the agenda correspond to possible meta-action executions. There is a close relationship to Blackboard[23] architectures. Meta-planning occurs when a planner constructs plans which involve explicit representations of meta-actions. See also Meta-level Inference[161].

Hayes-Roth, B. and Hayes-Roth, F., *A cognitive model of planning*, Cognitive Science **3** (1979), 275–310.

Stefik, M., *Planning and meta-planning (Molgen: part 2)*, Artificial Intelligence **16** (1981), 141–170.

Currie, K. and Tate, A., *BCS Expert Systems*, (British Computer Society Workshop Series) (Merry, M., ed.), Cambridge University Press, Cambridge, 1985, pp. 225–240.

Jim Doran

161 Meta-level Inference

The word 'Meta-X' can be read as 'X about X'. So, for instance the word 'Meta-knowledge' can be read as 'knowledge about knowledge', the word 'meta-level' as 'at a level about another level' and the word 'meta-level inference as 'inference performed at a level which is about another level'. More generally, when speaking of meta-level inference, we think of the following:

- There are two theories, one (usually called the *object theory*) which is about a given topic, and another one (usually called the *meta-theory*) which is about the object theory. The two theories may or may not in general share the same language.
- The inferences performed in the meta-theory are called 'meta-level inferences'. In general, the process of carrying on deduction in the meta-theory is called 'meta-level inference'.

For instance,

$$x^2 - 3x + 2 = 0 \quad \rightarrow \quad x = 1 \text{ or } 2.$$

$$x^2 - 3x + 2 = 0 \quad \text{contains two occurrences of the unknown '}x\text{'.}$$

are statements in respectively the theory and the meta-theory of algebra.

The goal of object-level inference is to obtain results in the topic of interest. The goal of meta-level inference is to obtain results about the theory of the topic of interest (the object theory) and then to use them to obtain *better results* about the topic of interest. Better results can mean 'to solve the problem in less time' (in which case the meta-level inference is used to control or guide the search at the object level), 'to get results otherwise unavailable' (in which case the meta-level inference is used to extend the solutions possibly obtainable by the object level) or 'to describe the object-level' (in which case the meta-level inference is used to get a better understanding on the object level which can then be used for further operations, e.g., learning), and so on.

Aiello, L. and Levi, G., *The uses of metaknowledge in AI sytems*, Proceedings of ECAI-84, 1984, pp. 705–717.

Fausto Giunchiglia, Frank van Harmelen, Robin Boswell

162 Meta-rules

In linguistics, meta-rules are a way of increasing the expressive power of a grammatical formalism. Informally, a grammar with meta-rules provides two ways of specifying rules: explicitly, by adding/specifying individual rules, or implicitly, by the mechanism of meta-rules. A meta-rule is a statement of the kind 'If the grammar contains a rule of a particular form, then the grammar also contains other rules whose form is systematically related to the form of the original rule.' Context-free Grammars[54] are well suited to extension by meta-rules. Meta-rules were extensively used in Generalised Phrase Structure Grammar (GPSG)[107]. For example, in GPSG the passive construction is added to the grammar by means of the passive meta-rule which uses the pattern of the active construction. Meta-rules are not transformations (as used in Transformational Grammar). They are also to be distinguished from derived rules which are special instances of other rules and are used in GPSG to implement long-distance dependencies.

Gazdar, G., *Phrase structure grammar*, The Nature of Syntactic Representations (Jacobson, P. and Pullum, G.K., eds.), D. Reidel Publishing Co., Dordrecht, 1981, pp. 131–186.

Hans Uszkoreit, Stanley Peters, *On some formal properties of meta-rules*, Linguistics and Philosophy **9** (1986), no. 4, 477–494.

Henry Thompson, Nicolas Nicolov

163 Minimal Window Search

(*Null Window Search*)

The use of minimal windows provides an improvement on the Alpha/Beta[8] tree-searching algorithm. Minimal window search is based on the assumption that all subtrees are inferior to the best subtree searched thus far, until proven otherwise. Having searched the first branch with a full width alpha-beta window [alpha,beta], all remaining branches are searched with a minimal window [alpha,alpha+1], where alpha represents the best Minimax[164] value found so far. If the value returned is indeed less than or equal to alpha, then the assumption was correct and the subtree is inferior. Otherwise, this subtree is superior and must be re-searched with a wider window to determine its exact minimax value.

Two things work in favour of minimal window search: First, it is easier to prove a subtree inferior than to determine its exact minimax value (like alpha-beta does), and second, most of the tests would result in exempting the subtree from further evaluation. This cautious 'testing-before-evaluating' strategy originally appeared in Pearl's Scout algorithm, and has subsequently been adapted to the negamax notation, giving the NegaScout, P-Alpha-Beta and PVS algorithms.

Pearl, J., *Heuristics: Intelligent Search Strategies for Computer Problem Solving*, Addison-Wesley, Reading, MA, 1984.

Alexander Reinefeld

164 Minimax

Consider a (partial) Game Tree[10] between two players. If the leaf nodes are labelled with payoffs such that the larger the value the more favourable the outcome for one of the players, then this player will naturally prefer the play to finish at a leaf node with as high a payoff as possible. We will refer to this player as Max, and to his opponent (who tries to direct the play to minimise Max's payoff) as Min. The Minimax value of such a tree is derived by starting at the leaf nodes and *backing up* values in the following manner. If the parent of a set of sibling nodes whose values, p_i, are all known is a Max node, then the value, P, of that node is given by $P=\max(p_i)$. If the parent is a Min node, then $P=\min(p_i)$. This procedure is recursive and results in eventually labelling the root node with a value. The optimal path through the tree from the point of view of both adversaries is known as the Maximin path.

Von Neumann, J. and Morgenstern, O., *The Theory of Games and Economic Behaviour*, John Wiley and Sons, New York, 1967.

Shannon, C.E., *Programming a computer for playing chess*, Philosophical Magazine **41**, 256–275.

Korf, R.E., *Search: a survey of recent results*, Exploring Artificial Intelligence (Survey Talks from the National Conferences on Artificial Intelligence) (Shrobe, H.E., ed.), Morgan Kaufmann, San Mateo, CA, 1988, pp. 197–237 (Chapter 6).

Hans Berliner

165 Modal Logic

True propositions can be divided into those—such as '2+2=4'—which are necessarily true and those—such as 'Mars has two moons'—which are contingently true. If 'α' is some proposition then we write '$\Box\alpha$' (or $L\alpha$) to mean 'α is necessarily true'. We can define '$\Diamond\alpha$' (or $M\alpha$), read as 'α is possibly true', as equivalent to '$\neg\Box\neg\alpha$'. '$\Box$' and '$\Diamond$' are modal operators. A modal logic is an extension of standard logic obtained by adding to some the modal operators together with appropriate axioms and rules of inference. A great many systems of modal logic have been constructed of which the most important are systems T, S4 and S5. The modal system T has as axioms some set of axioms appropriate to the standard propositional calculus together with:

$$\Box\alpha \rightarrow \alpha,$$

and

$$\Box(\alpha \rightarrow \beta) \rightarrow (\Box\alpha \rightarrow \Box\beta).$$

We add to the rules of inference of our standard logic a rule of necessitation to the effect that if α is theorem then so is $\Box\alpha$. The system S4 is obtained by adding to T:

$$\Box\alpha \rightarrow \Box\Box\alpha,$$

and S5 by adding to T:

$$\Diamond\alpha \rightarrow \Box\Diamond\alpha.$$

Modal predicate logic can be obtained by making analogous additions to standard Predicate Logic[208]. To construct a model of one of the above systems we postulate a set of possible worlds W, an assignment of a truth value to each atomic formula for each world and a relation of accessibility between a particular world and a subset of W. '$\Box\alpha$' is true in a world in such a model if and only if it is true in every world accessible to that world in that model. A proposition is valid if and only if it is true in every such model. If we require that the accessibility relation be reflexive then the valid propositions are those provable in T. If we also require that the system be transitive then we get as valid the theorems of S4. Finally if we also require that the relation be symmetric we get as valid the theorems of S5.

Hughes, G.E. and Cresswell, M.J., *An Introduction to Modal Logic*, Methuen, London, 1972.

Colin Phillips

166 Model-based Systems

A model-based system is the standard (or 'classical') way of building intelligent robot controllers. They are so-called because at the heart of the control system lies a representation or model of the task space (e.g., in a model-based navigation control system the representation is a map of the environment). Classical models, because they grew out of the example of a control engineer's plant model, generally contain very rich, accurate and detailed descriptions. Typically a model-based system will consist of a single monolithic controller which carries out what is called a *sense-think-act* cycle. In the first stage all the sensor data is read and compared with the model in order to assess the current state of the world. (Sophisticated systems may build their own models from sensor data, however it is more common for the model to be built by the programmer and given to the robot a priori.) Using symbolic planning techniques the control system then reasons about what actions must be performed in order to achieve its goals. The final part of this cycle involves actually carrying out the actions.

Newell, A., *The knowledge level*, Artificial Intelligence **18** (1982), 87–127.

Nitao, J.J. and Parodi, A.M., *Systems, Man and Cybernetics*, Proceedings of the 1985 IEEE International Conference, IEEE, New York, 1985, pp. 296–299.

Martin Westhead, Ashley Walker

167 Model-based Vision

A computer-vision technique in which interpretation of the image data is aided by some form of model of what is expected to be seen in the image. For example, a part inspection system would use a description (model) of where the key features of the part were and where specific tests should be undertaken. The main reasons for this approach are: (i) without some representation of the object no sensible identification is possible, (ii) the use of a geometric model may allow one to infer the position of the object in the scene and (iii) the model can be used to direct the image processing, thus saving considerable processing time.

The model may be implicit in the image analysis program, but more sophisticated approaches use explicitly defined models (allowing substitution of new models). The models may be expressed in many forms, including: geometric models (e.g., sizes and positional relationships between features), relational models (e.g., a graph with arcs representing pairwise relationships) or property models (e.g., a table of the expected properties, like area, colour, etc.).

There is a fuzzy dividing line between the use of 'model-based' for specific objects or object classes, and the use of the term for more generic features, such as lines. That is, a model of a line, or a curve, is a generally useful model and may be as precise and explicit as a model of a real object. In these cases, one often uses the term 'knowledge-based' because these features are more abstract, and their identification usually depends more on knowledge of how they appear in an image.

Fisher, R.B., *From Surfaces to Objects: Computer Vision and Three Dimensional Scene Analysis*, John Wiley and Sons, Chichester, 1989.

Robert B. Fisher

168 Model Matching

If an image contains an unknown object with f features, F_i of which are known to correspond to one of n known objects (models) M_j having m_j features, the problem of matching image features to model features can be 197zexpressed as an interpretation tree (IT). The root node of tree IT_j for model M_j has m_j descendants and the complete tree will have f levels. The total number of possible interpretations is therefore:

$$\sum_{j=1}^{n}(m_j)^f .$$

For even quite small numbers of features the interpretation tree quickly becomes too large to handle. Fortunately, very few interpretations in the tree are consistent with the data and by pruning those branches that are inconsistent the tree can be reduced to a more manageable size. The constraints used include the distances between features, the angles between features and the sign of the triple product between corresponding surface normals.

Grimson, W.E.L., *Object Recognition by Computer: The Role of Geometric Constraints*, MIT Press, Cambridge, MA, 1990.

H.W. Hughes

169 Modulation Transfer Function

Modulation transfer functions are used in pattern recognition as a description of a system's frequency response. It is a plot of the input to output amplitude ratio as a function of the frequency of the sinusoidally modulated input signal. See also Fourier Transform[97].

Campbell, F.W.C and Robson, J., *Application of Fourier analysis to the visibility of gratings*, J. Physiol. (Lond) **197** (1968), 551–566.

Jon Mayhew

170 Montague Semantics

A truth-conditional model-theoretical semantics for natural language, developed by Richard Montague, which attempts to formalise semantics as explicitly as the syntax usually is. The logic used is based on the Lambda Calculus[141] and uses the notion of possible worlds.

Dowty, D.R., Wall, R.E. and Peters, S., *Introduction to Montague Semantics*, D. Reidel Publishing Co., Dordrecht, 1981.

John Beavan

171 Morphographemics

(*Spelling Rules*)

The productivity of word-formation processes in many languages make it impossible to have every word in the lexicon. Systematic characterisation of the regularities in a computationally exploitable fashion is the concern of computational morphographemics. The characterisation has to cover both combination and re-spelling constraints on word components. Simple ad hoc approaches to segmenting words so that their components can be looked up in the dictionary are fairly easy to devise for languages like English.

Kay, M., *Morphological and syntactic analysis*, Linguistic Structures Processing (Zampolli, A., ed.), North-Holland, Amsterdam, 1977, pp. 131–234.

Koskenniemi, K., *Two-level model for morphological analysis*, Proceedings of IJCAI-83, vol. 2, 1983, pp. 683–685.

Koskenniemi, K., *Two-Level Morphology: A General Computational Model for Word-Form Recognition and Production*, Helsinki University, Helsinki, 1983.

Henry Thompson

172 Multi-actor System

(*Multi-agent system*)

A multi-actor system is a combination of two or more plan generation and plan execution systems (actors) which act and intercommunicate in a shared task environment. Communication actions between actors (e.g., request for information, provision of information, provision of partial plans) may be handled as ordinary actions and are closely related to speech acts in natural language theory. Techniques have been devised whereby one actor's beliefs about the beliefs of another may be expressed and computed.

Note: this use of the term 'actor' is substantially different from that employed in the actor model of computation (see Actors[6]).

Thorndyke, P.W., McArthur D. and Cammarata, S., *Autopilot: a distributed planner for airfleet control*, Proceedings of IJCAI-81, vol. 1, 1981, pp. 171–177, also in Readings in Distributed Artificial Intelligence (Bond, A.H. and Gasser, L., eds.) Morgan Kaufmann, San Mateo, CA, 1988, pp. 90–101.

Wilks, Y. and Bien, J., *Beliefs, point of view and multiple environments*, CSCM-6 (1981), Colchester Cognitive Studies Centre, University of Essex, also in Artificial and Human Intelligence (Elithorn, A. and Banerji, R., eds.) North-Holland, Amsterdam, 1984, pp. 147–171.

Jim Doran

173 Multi-layer Perceptrons

Multi-layer perceptrons are feed-forward networks built in a layered structure, with an input layer (directly linked to the environment), one or more hidden layers and an output layer for which there is desired pattern of activation, corresponding to a classification or prediction. Unlike single-layer Perceptrons[198], multi-layer perceptrons are capable of learning arbitrary input-output mappings. This is achieved through the use of hidden units, which have no direct link to the environment. Single-layer perceptrons may be trained trivially, since the weights in the network correspond directly to the input-output mapping under investigation. These dependencies are nested for multi-layer perceptrons and a means of adjusting the weights leading into hidden units is required. A solution to this credit assignment problem is the back-propagation of error algorithm, which propagates the required weight changes backwards through the network by employing the chain rule of differentiation.

Rumelhart D.E., Hinton, G.E. and Williams, R.J., *Learning internal representations by error propagation*, Parallel Distributed Processing (Rumelhart, D.E. and McClelland, J.L., eds.), vol. 1, MIT Press, Cambridge, MA, 1986, pp. 318–362.

Steve Renals

174 Multi-pulse LPC

(*Modified Linear Predictive Analysis, Synthesis of Speech, Vocoder*)

An improved technique of Linear Predictive Coding[149] in which the normal white noise or impulse excitation of the linear predictive filter is replaced by a modified impulse excitation only. A perceptually weighted spectrum matching technique is used to place the impulses.

Atal, B.S. and Remde, J.R., *A new model of LPC excitation for producing natural sounding speech at low bit rates*, Proceedings of the IEEE ICASSP, vol. 3, 1982, pp. 614–617.

Andrew Varga

175 Naive Physics

Naive physics identifies the core knowledge underlying commonsense physical intuition and allows the modelling of Qualitative Reasoning[223] in explaining or predicting changes in the physical world. It is a logic-based theory based on qualitative concepts of physics and the intuitive knowledge that lay people bring to bear in attempting to make sense of their physical world and in predicting significant events in the state of that world. For example, Hayes gives a theory for liquids in First-order Logic[208]; Allen and Kautz present a model of naive temporal reasoning for natural language understanding and problem solving. Rather than descriptions of systems in terms of algebraic formulae and differential equations, naive physics seeks to expose causal relationships and to reason about changes in physical systems. For instance, consider a physical system consisting of a bath with both taps running and the plug inserted. Without building a quantitative model in terms of rate of water flow, volume of the bath, etc. naive physics can predict that the carpet will (eventually) get wet.

Hayes, P., *The naive physics manifesto*, Formal Theories of the Commonsense World (Hobbs, J.R. and Moore, R.C., eds.), Ablex Pub. Corp., Norwood, 1985, also in Expert Systems in the Micro-electronic Age (Michie, D., ed.), Edinburgh University Press, 1979, pp. 242–270.

Helen Lowe

176 Natural Deduction

Natural deduction is a formal inference system which is said naturally to mirror the way in which humans reason. A natural deduction system consists of rules of inference eliminating and introducing each of the connectives and quantifiers of the Predicate Calculus[208]. There are twelve rules which may be used to infer conclusions. Two examples of such rules are given below:

$$\frac{A \qquad A \rightarrow B}{B} \qquad \frac{\begin{array}{c}\overline{A}\\ | \\ B\end{array}}{A \rightarrow B}$$

The rule on the left is that to eliminate the $\rightarrow$ connective and that on the right to introduce it. Proofs are trees, the leaves representing the assumptions, and the root, the conclusion that has been deduced from those assumptions. In natural deduction we can assume temporarily certain formulae, and then 'discharge' them later in the proof, using certain of the deduction rules. The $\rightarrow$ introduction rule above is an example of a rule discharging an assumption (A in this case). Natural deduction is sound and complete.

Prawitz, D., *Natural Deduction: A Proof Theoretical Study*, Almquist and Wiksell, Stockholm, 1965.

Tennant, N., *Natural Logic*, Edinburgh University Press, Edinburgh, 1978, reprinted with corrections, 1990.

Dave Plummer

177 Navigation

Robot navigation is a multi-faceted subject for which little general theory has been formulated, but many empirically derived solutions exist. Navigation systems can be built with elaborate model-based controllers or simple reactive behaviour based controllers (i.e., controllers using distributed, non-symbolic representations). The term is currently applied to mobile robots as well as assembly arms and usually involves a number of sensors (and sensor fusion techniques) and one or more actuating mechanisms. Traditionally, navigation systems were responsible for enabling the following behaviours: (i) obstacle avoidance, (ii) localisation (i.e., position determination), and (iii) path planning. Early research robots and many present-day industrial robots navigated with reference to an a priori specified map or a system of artificial beacons and signposts distributed through the environment. Recent interest in navigation systems for use in dynamic and unknown environments has caused the task of map building and maintenance to be considered as an essential navigation behaviour.

Chatila, R. and Laumond, J., *Position referencing and consistent world modelling for mobile robots*, Proceedings of the IEEE International Conference on Robotics and Automation, 1985, pp. 138–145.

Mataric, M.J., *Integration of representation into goal-driven behaviour based robots*, IEEE Transactions on Robotics and Automation, **8** (1992), 304–312.

Ashley Walker

178 Negation as Failure

Negation as failure is a rule of inference which allows one to deduce that NOT P is true if all possible proofs of P fail.

This is the way that negation is treated in Prolog[216] and Micro-PLANNER. When using a system for database access, the assumption that negation as failure corresponds to true negation is precisely a consequence of the closed world assumption, i.e., the assumption that all relevant information is contained in the database. Without the closed world assumption, negation as failure corresponds to 'we assume something is false if we cannot deduce it from available information', which is not the same as true negation. Negation as failure is an example of Default Reasoning[64]. Its mathematical meaning is extremely questionable.

Clark, K.L., *Negation as failure*, Logic and Databases (Gallaire, H. and Minker, J., eds.), Plenum Press, New York, 1978, pp. 293–322, also in Readings in Nonmonotonic Reasoning (Ginsberg, M.L., ed.) Morgan Kaufmann, Los Altos, CA, 1987, pp. 311–325.

Reiter, R., *On closed world databases*, Readings in AI and Databases (Mylopolous, J. and Brodie, M., eds.), Morgan Kaufmann, San Mateo, CA, 1989, pp. 248–258, also in Logic and Databases (Gallaire, G. and Minker, J., eds.) Plenum Press, New York, 1978, pp. 55–76.

Martin Merry

179 Neighbourhood Parallelism

The technique of using immediate parallel access to the whole set of locations that form the neighbourhood of an operation. This may be restricted to a single bit plane. See Cellular Arrays[36].

Danielsson, P. and Levialdi, S., *Computer architectures for pictorial information systems*, IEEE Computer (1981), 53–67.

T.P. Pridmore, S.R. Pollard, S.T. Stenton

180 Noise Reduction

(*Noise Smoothing*)

Imaging devices produce a certain amount of noise, that is, a random signal which does not carry any useful information. If the amplitude of the noise is higher than the amplitude of the signal that is to be analysed, the information cannot be directly extracted from the noisy signal. However, it is often the case that most of the noise energy is concentrated in high spatial frequencies (see Fourier Transforms[97]). If the signal to be extracted carries most of its information in lower spatial frequencies, it is possible to use a low pass filter (smoothing filter) which damps the higher frequencies. Then it is possible to extract the required information better.

Filtering is performed by convolving an image with a mask which is adapted to the task it should perform. Its shape can be determined exactly if the statistical distribution of the spectral components of signal and noise is known (Wiener filter). For simple low pass filters one often uses masks which approximate a

Gaussian in section. Simple masks like equal valued discs blur the image, but do not damp high frequencies as effectively as a Gaussian mask.

Pratt, W.K., *Digital Image Processing*, Wiley, New York, 1978.

Fritz Seytter

181 Non-linear Planning

(*Partially-ordered Plans, Parallel Planning*)

Non-linear planners are able to maintain the emerging plan as a partially-ordered network of actions. Unnecessary ordering (or linearisation) of the actions is avoided. Only when there are conflicts between parallel branches of the plan (such as the inability to determine a required condition at some point) is an ordering imposed. The first such system was Sacerdoti's NOAH.

A complete treatment of the handling of alternatives and all legal linearisations after an interaction between Subgoals[267] is detected was included in Tate's NONLIN. The ability to use the same technique in the presence of time constraints on particular actions was a feature of Vere's DEVISER. Most non-linear planners also use hierarchical planning techniques.

Sacerdoti, E.D., *A Structure for Plans and Behaviour*, Elsevier North-Holland, New York, 1977.

Tate, A., *Generating project networks*, Proceedings of IJCAI-77, vol. 2, 1977, pp. 888–893, also in Readings in Planning (Allen, J., Hendler, J. and Tate, A., eds.) Morgan Kaufmann, San Mateo, CA, 1990, pp. 291–296.

Vere, S.A., *Planning in time: windows and durations for activities and goals*, IEEE Trans. on Pattern Analysis and Machine Intelligence **5** (1983), no. 3, 246–267, also in Readings in Planning (Allen, J., Hendler, J. and Tate, A., eds.) Morgan Kaufmann, San Mateo, CA, 1990, pp. 297–318.

Austin Tate

182 Non-monotonic Reasoning

Non-monotonicity is a common feature of ordinary reasoning. For instance, if we are told that Tweety is a bird we assume Tweety can fly, but withdraw this when we are told that Tweety is a penguin.

Non-monotonic reasoning can be modelled using non-monotonic logics. Such logics differ from normal deductive logics in the effects of adding new axioms. In any type of logic, new axioms may give rise to new theorems, so that the set of theorems grows with the set of axioms. In non-monotonic logics, however, the set of theorems may *lose* as well as gain members when new axioms are added.

Default Reasoning[64] is a way of overcoming the problem of insufficient information; the system is told that unless it has information to the contrary, certain defaults are assumed to be true. Examples of systems utilising default reasoning are PLANNER and the Negation as Failure[178] of Prolog[216]. In the case of Prolog, any proposition which cannot be shown to be true from the statements in database is assumed to be false: this is the so-called closed world assumption. See also Truth Maintenance System[283], Default Logic[64], and Circumscription[39].

Special issue on non-monotonic logic, Artificial Intelligence **13** (1981), 1–172.
Ginsberg, M.L. (ed.), *Readings in Non-monotonic Reasoning*, Morgan Kaufmann, Los Altos, CA, 1987.

Luis Jenkins, Dave Plummer

183 Numerically-controlled Machine Tools

A machine tool is a power-driven machine for cutting, grinding, drilling, or otherwise shaping a metal workpiece. A numerically-controlled (NC) machine tool is a system in which actions are controlled by numerical data given by a program. The program can position a tool point in three dimensions relative to a workpiece, and can control other functions such as speed, coolant flow, and selection of tools. Most NC tools can be programmed in the APT language with FORTRAN[95] subroutines. The capability to perform 3-D geometrical calculations makes it possible to sculpt arbitrarily complicated shapes. Sensory Feedback[247] is used to report differences between actual and desired tool movements, and between actual and desired workpiece dimensions to cause the NC tool to correct or minimise the error. This can automatically compensate for tool wear and detect tool breakage. Robot manipulators can be used to feed blank workpieces and to remove finished ones.

Simon, W., *The Numerical Control of Machine Tools*, Edward Arnold, 1978.
Faux, I.D. and Pratt, M.J., *Computational Geometry for Design and Manufacture*, Ellis Horwood, Chichester, 1979.

W.F. Clocksin

184 Object-centred Co-ordinates

An object is described in object-centred co-ordinates if the coordinate origin is fixed with respect to the object, the coordinate axes are defined by the structure of the object, and all other object points are defined with respect to that system. This description is, then, independent of viewpoint. One task of a vision system is to move from Viewer-centred Description[291] to object-centred description.

Marr, D., *Vision: A Computational Investigation into the Human Representation and Processing of Visual Information*, W.H. Freeman, San Francisco, CA, 1982.
Martin, S.J., *Numerical Control of Machine Tools*, Hodder and Stoughton, London, 1970.

T.P. Pridmore, S.R. Pollard, S.T. Stenton

185 Object-oriented Programming

(OOP)

Object-oriented programming dates back to the 1960s; it developed in response to software engineers' needs to handle function overloading, data abstraction and code re-use in a clean way. In object-oriented programming one typically defines a set of classes, arranged in an inheritance hierarchy or lattice. A class is a definition; instances conforming to that definition can then be defined. Functions are attached to classes. Thus, for example, addition may be independently defined for strings and for numbers; typically a compiler or interpreter then sorts out which function is to be invoked for the expression 'A+B', by considering the type of A, and in some systems also of B. A jargon has evolved with the technology: for example, a 'method' is a function definition, a 'message' is a function call. Free variables in function definitions refer to variables named in the class definition to which the function is attached, or in any class from which that class inherits. There are many extensions to the basic idea sketched here.

Object-oriented programming is of clear relevance to AI, not only because of AI's software engineering requirements but also from a knowledge representation perspective. KRL (qv) and its descendants have made a point of keeping data and procedures together in an object-oriented programming-like way. Many AI theories can be expressed fairly cleanly in object-oriented programming since the meaning of a message is determined by the receiver rather than by the sender. C++ is perhaps the best-known object-oriented programming language generally, CLOS (Common Lisp Object System) the most used within AI, and Smalltalk the oldest and purest. Most versions of Lisp[150], Prolog[216] and Scheme now have object-oriented programming features built in or available as extensions.

Meyer, Bertrand, *Object-oriented Software Construction*, Prentice-Hall, Englewood Cliffs, NJ, 1988.

Keene, Sonya E., *Object-oriented Programming in COMMON LISP: A Programmer's Guide to CLOS*, Addison-Wesley, Reading, MA, 1989.

Peter Ross

186 One-Then-Best Backtracking

Since there is often good local information available to indicate the preferred solution path, it is often best to try out that indicated by heuristic information before considering the many alternatives that may be available should the local choice prove faulty. Taken to the extreme, Depth-first Search[71] gives something of the flavour of such a search strategy. However, gradual wandering from a valid solution path could entail backtracking through many levels when a failure is detected.

An alternative is to focus on the choice currently being made and try to select one of the local choices which seems most promising. This continues while all is going well (perhaps with some cut-off points to take a long, hard look at how well things are going). However, if a failure occurs, the *whole* set of alternatives which have been generated (and ranked by a heuristic evaluator) are reconsidered to select a re-focussing point for the search. Such a search strategy is used in Tate's NONLIN, for example.

Adeli, H. (ed.), *Knowledge Engineering, Volume 1: Fundamentals*, McGraw-Hill, New York, pp. 157–199.

Austin Tate

187 Operators

(*Operator Schema, Action Schema*)

In the context of planning, operators represent the (generic) actions which the actor can execute in the planning environment. Operators are commonly represented as three lists: a list of preconditions, a list of facts that will be true after application of the operator (add list) and a list of facts that will no longer be true after the operator is applied (delete list). These lists normally involve variables which must be bound for any particular instance of the action.

The difficulty encountered in fully capturing the effects of actions in non-trivial planning domains using such operators is an aspect of the frame problem.

Nilsson, N.J., *Principles of Artificial Intelligence*, Tioga Pub. Co., Palo Alto, CA, 1980, and Springer-Verlag, Berlin, 1982.

Maarten van Someren, Jim Doran

188 Operator Tables

(*Triangle Table*)

When Macro Operators[155] are built from elementary Operators[187], the original operators and the macro operator are available, but not any sub-sequences of the macro operator. For example, if O(20) was built from O(1) through O(10), then it is not possible to use the sub-sequence O(8) through O(10). The representation in the figure accompanying this entry was devised to overcome this inefficiency. The preconditions of each operator are in the row preceding it, while the postconditions are placed in the columns below. The execution of subsequent operators may affect these postconditions, so they are updated in subsequent rows of the table by considering each operator's delete list.

O1's Preconditions	operator O1		
O2's preconds not established by O1	postconds of O1	operator O2	
O3's preconds not established by O1 or O2	facts in cell above minus delete O2	postconds of O2	operator O3
	facts in cell above minus delete O3	facts in cell above minus delete O3	postconds of O3

Now, if we define the nth kernel of the table as the intersection of all rows below and including the nth and all columns to the left and including the nth, we can easily verify whether the ith action in an operator table is currently executable by checking whether all the predicates contained in the ith kernel are true. Giving a planner a library of operator tables can reduce the search on new problems (see Library-based Planning[147]). However, the entries in the tables should be generalised (with respect to variable bindings) if they are to be of any use.

Fikes, R.E., Hart, P.E. and Nilsson, N.J., *Learning and executing generalized robot plans*, Artificial Intelligence **3** (1972), 251–288, also in Readings in Planning (Allen,

J., Hendler, J. and Tate, A., eds.) Morgan Kaufmann, San Mateo, CA, 1990, pp. 189–206.

Maarten van Someren

189 Opportunistic Search

Some systems do not have a fixed (Goal-driven[16] or Data-driven[61]) directional approach to solving a problem. Instead a current 'focus' for the search is identified on the basis of the most constrained way forwards. This may be suggested by comparison of the current goals with the initial world model state, by consideration of the number of likely outcomes of making a selection, by the degree to which goals are instantiated, etc. Any problem-solving component may summarise its requirements for the solution as constraints on possible solutions or restrictions of the values of variables representing objects being manipulated. It can then suspend its operations until further information becomes available on which a more definite choice can be made.

Many such systems operate with a Blackboard[23] through which the various components can communicate via constraint information. The scheduling of the various tasks associated with arriving at a solution may also be dealt with through the blackboard.

Hayes, P.J., *A representation for robot plans*, Proceedings of IJCAI-75, 1975, pp. 181–188, also in Readings in Planning (Allen, J., Hendler, J. and Tate, A., eds.), Morgan Kaufmann, San Mateo, CA, 1990, pp. 154–161.

Hayes-Roth, B. and Hayes-Roth, F., *A cognitive model of planning*, Cognitive Science **3** (1979), 275–310.

Stallman, R.M. and Sussman, G.J., *Forward-reasoning and dependency-directed backtracking*, Artificial Intelligence **9** (1977), 135–196.

Austin Tate

190 Optical Flow

The instantaneous positional relative velocity field of a scene projected onto an imaginary surface: a retinal velocity map. Mathematical analysis has shown that from a monocular view of a rigid, textured, curved surface it is possible to determine the gradient of the surface at any point and the motion relative to that surface from the velocity field of the changing retinal image and its first and second spatial derivatives. See Intrinsic Images[134].

Longuet-Higgens, H.C. and Prazdny, K., *The interpretation of moving retinal images*, Proceedings of the Royal Society, London **B208** (1980), 385–397.

T.P. Pridmore, S.R. Pollard, S.T. Stenton

191 Optimistic Plan Execution

When a Plan Structure[202] is executed, there is no guarantee that the sequence of planning environment states predicted by the plan will actually occur. Deviation from expectation due to faulty planning or faulty Operator[187] execution is very likely.

When an optimistic plan execution technique executes an operator in a plan and so encounters a new environment state, it attempts to identify the new state by matching it successively against all the states predicted in the plan, working backwards from the goal. The first match found determines which operator in the plan is to be executed next. If no match is found, then re-planning is needed.

The advantage of this somewhat laborious technique is that the execution of unnecessary operators may be avoided and some execution failures can be overcome by repeating the execution of operators.

Fikes, R.E., Hart, P.E. and Nilsson, N.J., *Learning and executing generalized robot plans*, Artificial Intelligence **3** (1972), 251–288, also in Readings in Planning (Allen, J., Hendler, J. and Tate, A., eds.) Morgan Kaufmann, San Mateo, CA, 1990, pp. 189–206.

Jim Doran

192 Paramodulation

A rule of inference of Predicate Calculus[208], used to deduce a new formula from two old ones. It is used in Automatic Theorem Proving[276], in conjunction with Resolution[234], as an alternative to the axioms of equality. All the three formulae involved must be in Clausal Form[41]. If C[t'] and D are clauses, where C[t'] indicates that C contains a particular occurrence of t', then the rule is:

$$\frac{C[t'] \qquad D \vee s = t}{(C[s] \vee D)\theta} \qquad \frac{C[t'] \qquad D \vee t = s}{(C[s] \vee D)\theta},$$

where θ is the most general unifier of t and t', and is obtained by Unification[286]. C[s] indicates that only the particular occurrence of t' is replaced by s in C, other occurrences of t' are untouched.

Chang, C. and Lee, R.C., *Symbolic Logic and Mechanical Theorem Proving*, Academic Press, New York, 1973.

Bundy, A., *The Computer Modelling of Mathematical Reasoning*, Academic Press, London, 1983, pp. 80–83, also second edition, 1986.

Alan Bundy

193 Partial Evaluation

(*Partial Deduction*)

The main goal of partial evaluation is to perform as much of the computation in a program as possible by using a partial specification of the input values of the program. The theoretical foundation for partial evaluation is Kleene's S-M-N theorem from Recursive Function Theory[225].

The partial evaluation algorithm takes as its input a function (program) definition, together with a partial specification of the input of the program, and produces a new version of the program that is specialised for the particular input values. The new version of the program will then be less general but more efficient then the original version.

The partial evaluation algorithm works by symbolically evaluating the input program while in the mean-time trying to: (i) propagate constant values through the program code, (ii) unfold procedure calls, and (iii) branch out conditional parts of the code. If the input program is logical, then the symbolic evaluation of the program becomes the construction of the proof tree corresponding to the execution of the program.

In recent years partial evaluation in Logic Programming[153] has attracted particular attention in connection with meta-programming as an optimisation technique. See also Program Transformation[214].

van Harmelen, F., *The limitations of partial evaluation*, Logic-based Knowledge Representation (Jackson, P., Reichgelt, H, van Harmelen, F., eds.), MIT Press, Cambridge, MA, 1989, pp. 87–111 (Chapter 5).

Lakhotia, A. and Sterling, L., *ProMix: a Prolog partial evaluator system*, The Practice of Prolog (Sterling, L., ed.), MIT Press, Cambridge, MA, 1990, pp. 137–179.

Frank van Harmelen

194 Partitioned Semantic Net

Means of enhancing the organisational and expressive power of Semantic Nets[244] through the grouping of nodes and links, associated with Hendrix. Nodes and links may figure in one or more 'spaces', which may themselves be bundled into higher-level 'vistas', which can be exploited autonomously and structured hierarchically. The effective encoding of logical statements involving connectives and quantifiers was an important motivation for partitioning, but the partitioning mechanisms involved are sufficiently well-founded, general and powerful to support the dynamic representation of a wide range of language and world knowledge; and partitioned nets have been extensively used for a range of such purposes at SRI.

Hendrix, G.G., *Encoding knowledge in partitioned networks*, Associative Networks (Findler, N.V., ed.), Academic Press, New York, 1979, pp. 51–92.

Karen Sparck Jones

195 Pattern Directed Languages

This class of languages is Data-driven[61]. That is, the current context is used to determine which procedure to call next—sometimes called Pattern Directed Invocation[196]. This is opposed to the fixed sequence of procedure calls in languages like FORTRAN[95], Algol etc. There is a corresponding necessity for suitable Pattern Matching[197] to take place.

Example languages include PLANNER (and Micro-PLANNER), CONNIVER, QLisp (and QA4), Prolog[216] and Production Rule Systems[illegible] like OPS5.

Different styles of reasoning are possible: OPS5, for example, uses Forward Chaining[96] while Prolog uses Backwards Chaining[16]. Also, different styles of pattern matching are used: PLANNER uses an extremely simple algorithm while QLisp can match pattern variables to arbitrary Lisp[150] structures.

The problem of conflict resolution also arises when, in principle, one pattern is suitable for several different procedure invocations. This is solved in Prolog by using a fixed search strategy and in OPS5 with a simple conflict resolution algorithm.

Waterman, D.A. and Hayes-Roth, F. (eds.), *Pattern-directed Inference Systems*, Academic Press, New York, 1978.

Paul Brna

196 Pattern Directed Retrieval/Invocation

The retrieval of a datum from a database by virtue of its syntactic form. A pattern (or template) of the required datum is compared with the data in the database; data that 'match' with the pattern are retrieved. A successful match may bind the unspecified parts (variables) of either or both of the pattern and 'target' datum. In the case of procedure invocation, a pattern is associated with each procedure. Calls to a procedure are made when the current situation or goal matches the associated pattern of the procedure. This allows for a more flexible control flow as procedures are not called by name but by content. This form of retrieval/invocation is central to the PLANNER type languages; two way Pattern Matching[197] (or Unification[286]) being the basis of the Resolution[234] principle on which Prolog[216] is based.

Waterman, D.A. and Hayes-Roth, F. (eds.), *Pattern-directed Inference Systems*, Academic Press, New York, 1978.

Lincoln Wallen

197 Pattern Matching

In its most general form a pervasive feature of, or basis for, AI systems: the essential objective is to test whether a specific received data structure is an instance of a given general pattern, and particularly, to establish whether input data items can provide values for pattern variables. The matching can be made more or less fuzzy, according to the conditions on the individual pattern variables, and on their joint satisfaction. Pattern matching is important in AI because it reflects (i) the fact that complex general concepts exhibit considerable variation in particular manifestations, and (ii) the fact that individual elements of these participate in relationships with one another. See also Unification[286].

Nilsson, N.J., *Principles of Artificial Intelligence*, Tioga Pub. Co., Palo Alto, CA, 1980, and Springer-Verlag, Berlin, 1982.

Karen Sparck Jones

198 Perceptrons

A parallel decision making mechanism that superficially resembles the sort of processing that may be characteristic of neurons in the brain. It is a pattern recognition device with a threshold. If the linear combination of the 'weighted inputs' is greater than some threshold value then the perceptron 'fires'. It is possible for a perceptron to learn, e.g., if the weights associated with the inputs that were active in the case of a false alarm are decreased, and weights associated with inputs that were active in the case of a miss are increased then it is intuitively plausible that recognition performance will improve. There is a theorem that says that a perceptron will learn to recognise a class correctly over a finite number of errors. The analysis of the mathematical properties of perceptrons revealed profound limitations to their competence. These limitations are largely due to the difficulties inherent in making global decisions on the basis of only local evidence. Thus a perceptron can't tell whether a figure is connected or not, or whether there is one, and only one, instance of a pattern present. Nevertheless there is a resurgence of interest in perceptrons associated with the development of connectionist schemes for visual processing. See Connectionism[48], and Multi-layer Perceptrons[173].

Hertz, J., Krough, A. and Palmer, R.G., *Introduction to the Theory of Neural Computation*, Addison-Wesley, Redwood City, CA, 1991.

Minsky, M. and Papert, S., *Perceptrons*, MIT Press, Cambridge, MA, 1969.

Jon Mayhew

199 Photometric Stereo

A technique for recovering shape information from Reflectance Maps[227]. Given an image and a reflectance map for a light source, the intensity at a particular image location constrains the surface orientation to lie on a contour in the reflectance map. If three images are taken with different light source positions the surface orientation must lie on a known contour in each of the three associated reflectance maps, so the intersection specifies the surface orientation.

Marr, D., *Vision: A Computational Investigation into the Human Representation and Processing of Visual Information*, W.H. Freeman, San Francisco, CA, 1982.

T.P. Pridmore, S.R. Pollard, S.T. Stenton

200 Pitch Extraction

There are three main algorithms used for performing pitch extraction for speech synthesis and recognition: the Gold-Rabiner parallel processing method with voiced-unvoiced decision, the Dubnowski autocorrelation method and the Harmonic-sieve pitch extraction method.

Gold, B. and Rabiner, L., *Parallel processing techniques for estimating pitch periods of speech in the time domain*, The Journal of the Acoustical Society of America **46** (1969), no. 2, 442–448.

Gold, B., *Note on buzz-hiss detection*, The Journal of the Acoustical Society of America **36** (1964), no. 9, 1659–1661.

Dubnowski, J.J., Schafer, R.W. and Rabiner, L.R., *Real-time digital hardware pitch detector*, IEEE Trans. ASSP **24** (1976), 1–8.

Duifhuis, H., Willems, L.F. and Sluyter, R.J., *Measurement of pitch in speech: an implementation of Goldstein's theory of pitch perception*, JASA **71** (1982), no. 6, 1568–1580.

Andrej Ljolje

201 Plan Recognition

(*Plan Inference*)

Given a sequence of actions carried out in a task domain by some actor, it is possible to draw conclusions about the plan that the actor is executing by working bottom up. If the actor's top goal is known or can be guessed, then a top down matching process is feasible. The task of plan recognition can be related to Chart Parsing[38] and can be partially formalised using a first order theory of action and Circumscription[39].

Schmidt, C.F., Sridharan, N.S. and Goodson, J.L., *The plan recognition problem*, Artificial Intelligence **11** (1978), 45–83.

Allen, J. and Perrault, C., *Analyzing intention in utterances*, Artificial Intelligence **15** (1980), 143–178.

Jim Doran

202 Plan Structure

A plan structure embodies a partially or totally ordered set of partially instantiated Operators[187] together with the (sub) goals to which they are directed, and information which indicates how the planning environment is predicted to change as the plan is executed. Plan structure can be contrasted with Goal Structure[109], as used in Operator Tables[188] where the purposes of the plan's parts are recorded.

Nilsson, N.J., *Principles of Artificial Intelligence*, Tioga Pub. Co., Palo Alto, CA, 1980, and Springer-Verlag, Berlin, 1982.

Jim Doran

203 POPLOG

POPLOG is a multi-language AI software development environment developed mainly at the University of Sussex, UK. Currently, POP-11, Prolog[216], Common Lisp[150], and Standard ML[258] are all supported, each with its own libraries, help information and editor customisations. POPLOG comes with an interface to X and facilities for loading procedures written in other languages, such as C. Many useful AI packages, such as neural net simulators and expert system shells, are also provided, implemented in POPLOG. There is a POPLOG user group and a POPLOG electronic mail forum.

Historically, POPLOG developed from the programming language POP-2 of Burstall, Collins and Popplestone, which was devised in Edinburgh around 1970. POPLOG started as an incremental compiler for a modified language, POP-11, which combined the flexibility of working with an interpreter and the efficiency of compiled code. This was combined with an integrated programmable screen editor and extensive help and library facilities.

A vital part of the POPLOG architecture is the use of a virtual machine; POPLOG code is compiled into instructions for the POPLOG virtual machine, which are then translated into native code for the hardware it is running on. Procedures to generate virtual machine instructions are available to the user, which means that it is quite straightforward in POPLOG for a user to write a compiler for a new language which produces efficient code. This was the means by which the multi-language nature of POPLOG was realised, and now compilers for POP-11, Prolog, Common Lisp and Standard ML are distributed as standard with the system. These languages come with their own development environments, but the POPLOG system is relatively compact, compared for instance to Lisp systems of similar complexity. Because the languages share a model of computation at the level of the POPLOG virtual machine, flexible mechanisms for combining programs written in different languages can be provided. This design also means that POPLOG can be ported to new platforms relatively quickly.

Anderson, J.A.D.W. (ed.), *POP-11 Comes of Age: The Advancement of an AI Programming Language*, Ellis Horwood, Chichester, 1989.

Chris Mellish

204 Possibilistic Logic

Possibilistic logics assign degrees of necessity and possibility to sentences, which express the extent to which these sentences are believed to be necessarily or possibly true, respectively. The semantics of possibilistic logics is defined based on a Fuzzy Set[103] of possible worlds, i.e., a set of possible worlds together with a fuzzy membership function that describes the extent to which each possible world belongs to a referential set. Given a sentence a, the *necessity measure* $N(a)$ expresses the greatest lower bound for the degree of membership of the worlds that support a, and the related *possibility measure* $\Pi(a)$ expresses the least upper bound for this degree of membership.

Possibilistic logics should not be confused with Fuzzy Logics[102]. Although the concept of fuzzy measures is central to the description of both of them, in possibilistic logics the truth-values of sentences range on $\{true, false\}$ only and necessity and possibility measures describe states of belief, whereas in fuzzy logics the truth-values of sentences range on the interval $[0, 1]$ and fuzzy measures describe 'degrees of trueness'.

Sombé, L., *Reasoning under incomplete information in artificial intelligence: a comparison of formalisms using a single example*, International Journal of Intelligent Systems **5** (1990), no. 4, 323–472.

Flávio Corrêa da Silva

205 Postulating Intrinsic Properties

Given two symbolic independent variables X and Y, and one numeric dependent variable Z, one cannot find a numeric law relating these terms (since only one is numeric). However, suppose that by holding X constant and varying Y, one observes different values for Z. Then one can postulate an intrinsic property I whose values are set to those of Z, and one can associate these numeric values with the symbolic values of Y. Since one now has two numeric terms Z and I, one can find the law $Z/I = 1.0$ (which is true tautologically). However, when a new value of X is examined and the same values of Y are used, one can retrieve the intrinsic values associated with Y. In this new context the ratio Z/I may take on some other constant value, which is an empirically meaningful law. This method can be used to infer intrinsic properties such as mass, specific heat, and the index of refraction from combinations of symbolic and numeric observations.

Langley, P., Bradshaw, G. and Simon, H.A., *Rediscovering chemistry with the Bacon system*, Machine Learning: An Artificial Intelligence View (Michalski, R., Carbonell, J. and Mitchell, T., eds.), Tioga Pub. Co., Palo Alto, CA, 1983, pp. 307–330.

Bradshaw, G., Langley, P. and Simon, H.A., *Bacon.4: the discovery of intrinsic properties*, Proceedings of the Third National Conference of the Canadian Society for Computational Studies of Intelligence, 1980, pp. 19–25.

Pat Langley

206 Precondition Analysis

Precondition analysis is an analytic strategy-learning technique. Precondition analysis operates in two phases: the learning cycle and the performance cycle. In the learning cycle, the program is given an example of a correctly executed task.

The example may contain several individual steps, each step being the application of an Operator[187]. The program first examines the example, to find out which operators were used in performing the task. This stage is called Operator Identification. During this phase, the program may discover that it doesn't possess the relevant operator, and the user is asked to provide the necessary information.

Once this phase is complete the program builds an explanation of then strategic reasons for each step of the task. The explanation is in terms of satisfying the preconditions of following steps.

From this explanation, it builds a plan that is used by the performance element. These plans are called schemas. The performance element executes the schemas in a flexible way, using the explanation to guide it. The explanations are used to make sensible patches if the plan can't be used directly.

Precondition analysis has been implemented in LP (Learning PRESS), a program that learns new techniques for solving algebraic equations. Precondition analysis is somewhat similar to the explanation-based generalisation (see Explanation-based Learning[88]) approach of Mitchell, but differs in that precondition analysis can work in domains in which the methods are not invertible, and in situations where the domain theory is imperfect.

Silver, B., *Learning equation solving methods from examples*, Proceedings of IJCAI-83, vol. 1, 1983, pp. 429–431.

Silver, B., *Precondition analysis: learning control information*, Machine Learning: An Artificial Intelligence Approach (Michalski, R.S., Carbonell, J.G. and Mitchell, T.M., eds.), vol. 2, Morgan Kaufmann, Los Altos, CA, 1986, pp. 647–670 (Chapter 22).

Bernard Silver

207 Predictive Parsing

(*Expectation-based Parsing*)

Predictive parsing is an approach to natural language analysis based on the use of powerful programs associated primarily with individual lexical items which embody expectations about the form and content of the subsequent sentence/text input. The expectations are intended to determine the analysis of further inputs by providing semantic structures into which they must fit. (These structures may be those of Preference Semantics[209], Conceptual Dependency[45], or, by generalisation, Scripts[239].) The essentially word-driven basis, and emphasis on semantic rather than syntactic information and processing, of this style of parsing makes it predictive in a very different sense from that of classical top-down syntactic parsing; and allowing the exploitation of any type or piece of information in a word program offers great flexibility. The price includes a lack of generality, and overlaps or gaps between individual programs. The approach is therefore best suited to domain-limited language processing.

The main application of expectation-based parsing has been in the extraction of conceptual dependency structures from English; the technique has also been generalised in the direct application of scripts to parsing. Word expert parsing is a specialisation of this approach confined to word-based expectations, but not restricted to left-to-right operation.

Birnbaum, L. and Selfridge, M., *Conceptual analysis of natural language*, Inside Computer Understanding (Schank, R.C. and Riesbeck, C.K., eds.), Lawrence Erlbaum Associates, Hillsdale, NJ, 1981, pp. 318–353.

Karen Sparck Jones

208 Predicate Calculus

(*Predicate Logic, First-order Logic*)

Predicate calculus is a formal language in which it is possible to express statements about simple domains. It comprises a set of symbols, and rules for combining these into terms and formulae. There are also rules of inference, which state how a new formula can be derived from old formulae. A logical system will have an initial set of sentences ('axioms') and any sentence which can be derived from these axioms using the inference rules is called a 'theorem' of the system.

The standard logic is a well-defined notation in which exact descriptive statements can be formulated about any 'model' (i.e., set of objects with relations between them). Moreover, purely formal manipulations of these symbolic statements (i.e., inference) can be used to produce further valid descriptions of that same model, without direct reference to the model itself.

Terms can be constants (names of objects), variables (marking which part of a formula is quantified) or functions applied to arguments (e.g., $f(a,b,x)$). Atomic sentences are formed by applying a predicate to a set argument term (e.g., $\mathrm{P}(f(a,b,x),c,g(h(y)))$). Compound sentences are formed by adding negation to a sentence (e.g., $\neg\mathrm{R}(a,b)$), joining two statements with a connective such as $\wedge$ ('and'), $\vee$ ('or'), $\rightarrow$ ('implication'). There are two quantifiers which, used together with variables, allow the expression of universal statements, such as

$$(\forall x)Q(x)\text{: 'for all x, Q(x)'},$$

and existential statements:

$$(\exists z)R(z,a)\text{: 'there exists a z such that R(z,a)'}.$$

Automatic inference techniques (e.g., Resolution[234]) for first-order logic have been widely explored.

Smullyan, R., *Mathematical Logic*, Springer-Verlag, Berlin, 1968.
Crossley, J.N. et al., *What is Mathematical Logic?*, Oxford University Press, Oxford, 1972.

Graeme Ritchie

209 Preference Semantics

An approach to language understanding most fully developed by Wilks. Preference semantics recognises that, in general, semantic constraints on word combinations cannot be absolute, as this would be incompatible with the creativity of language. The semantic patterns embodied, e.g., in Case Frames[33], and expressed, e.g., by Semantic Primitives[245] indicate the mutual contextual preferences of words: for example 'hit' meaning STRIKE prefers a HUMAN agent. Wilkes' special contribution was a preference metric, minimising information in the interpretation selected. Thus word sense and sentence structure selection in text processing is then determined by maximum Constraint Satisfaction[52], and does not depend on complete satisfaction. Analogously, where interpretation involves inference, the metric selects the interpretation depending on the shortest inference chain.

Wilks, Y.A., *A preferential, pattern seeking semantics for natural language inference*, Artificial Intelligence **6** (1975), 53–74.

Karen Sparck Jones

210 Primal Sketch

A term used by Marr for a representation making explicit the properties of intensity changes in the retinal image(s). The raw primal sketch makes explicit only very localised properties, such as size, position and orientation; the full primal sketch, resulting from grouping elements of the raw primal sketch, makes explicit more global properties such as alignment. Marr claimed that higher level processes interact only with the primal sketch and its derivatives, not with the data from which the primal sketch is derived. The primal sketch is computed by dedicated processors which are independent of higher level processes.

Marr, D., *Vision: A Computational Investigation into the Human Representation and Processing of Visual Information*, W.H. Freeman, San Francisco, CA, 1982.

T.P. Pridmore, S.R. Pollard, S.T. Stenton

211 Principle of Least Commitment

This principle asserts that decisions should be deferred as long as possible before being taken, so that when they are taken the probability of their correctness is maximised. In the planning context, the principle is used to justify deferring action ordering decisions, and object choice decisions during plan elaboration. A consequence is an emphasis on the posting and propagation of Constraints[52] that an object must satisfy. See also Delayed Evaluation[66] and Lazy Evaluation[143].

Sacerdoti, E.D., *A Structure for Plans and Behaviour*, Elsevier North-Holland, New York, 1977.

Stefik, M.J., *Planning with constraints (Molgen: part 1)*, Artificial Intelligence **16** (1981), no. 2, 111–140, also in Readings in Planning (Allen, J., Hendler, J. and Tate, A., eds.) Morgan Kaufmann, San Mateo, CA, 1990, pp. 171–185.

Jim Doran

212 Procedural Attachment

(*Procedural Embedding*)

It is often necessary in knowledge representation for an executable procedure to be directly associated with one or more data structures, in order to indicate when it should be used. Attached procedures lie dormant until certain conditions are satisfied, when they execute; e.g., in KRL two types of attached procedures are identified (which hold for most systems):

- servants: these are executed when some operation should be applied to a data object (or set of data objects). A selection mechanism may be required to select the right procedure if more than one is available.
- Demons[67]: these are invoked when something has been done or is about to be done, (note that all demons whose conditions are met are activated).

Two methods of attachment were identified:

- traps: these are attached to individual data objects and apply to operations and events involving the unit to which they are attached.
- triggers: these are attached to classes of data objects and apply to operations and events involving any objects in the class to which they are attached.

Both traps and triggers may be either servants or demons (cf. 'methods' in CONNIVER and 'theorems' in Micro-PLANNER).

Bobrow, D.G. and Winograd, T., *An overview of KRL, a knowledge representation language*, Cognitive Science **1** (1977), 3–46, also in Readings in Knowledge Representation (Brachman, R.J. and Levesque, H.J., eds.), Morgan Kaufmann, Los Altos, CA, 1985, pp. 263–285.

Robert Corlett

213 Production Rule System

(*Production System, If/Then Rules, Situation/Action Rules*)

A programming language where the programs are condition $\Longrightarrow$ action rules. These are interpreted by the following operations: all rules whose conditions are satisfied are found, one of them is selected, and its action is called, the process is then repeated. They have been used extensively in computational psychology and knowledge engineering.

Newell, A. and Simon, H.A., *Human Problem Solving*, Prentice-Hall, Englewood Cliffs, NJ, 1972.

Luger, G.F. and Stubblefield, W.A., *Artificial Intelligence: Structures and Strategies for Complex Problem Solving*, Benjamin/Cummings Publishing Company, Redwood City, CA, 1993, pp. 163–179.

Alan Bundy

214 Program Transformation

A technique for developing programs. An initial specification is written as a (probably inefficient) program and then transformed to an efficient version using methods guaranteed to preserve the meaning of the program. Within the declarative languages program transformations can be based on a small set of provably correct basic transformations facilitating the development of semi-automatic transformation systems. See also Partial Evaluation[193].

Darlington, J., Henderson, P. and Turner, D.A. (eds.), *Functional Programming and its Applications: An Advanced Course*, Cambridge University Press, Cambridge, 1982, pp. 193–215.

John Darlington

215 Programming Cliche

(*Idiom*)

Inspection methods are problem solving techniques which rely upon recognising the form of a solution. In programming, there is a set of standard forms from which a wide range of programs can be constructed. This set of standard forms includes such things as particular control strategies with unspecified primitive actions. Such standard forms are know as programming cliches. The term originates from the MIT Programmer's Apprentice project. As a result of being more abstract, programming cliches are much more adaptable concepts than macros or subroutines.

Rich, C. and Waters, R.C., *The Programmer's Apprentice*, ACM Press and Addison-Wesley, New York, 1990.

Perlis, A.J. and Rugaber, S., *Programming with idioms in APL*, Proceedings of APL79, 1979.

Kevin Poulter

216 Prolog

Prolog is a programming language originally developed by Colmerauer around 1972. It can be seen as an attempt at a pure Logic Programming[153] language as it is based on the Clausal Form[41] of first order predicate logic restricted to Horn clauses. Essentially, it is a Resolution Theorem Prover[234].

Since the influential implementation of Prolog by D.H. Warren in 1976 and his development of the Warren Abstract Machine (WAM), Prolog has become more suitable for large programming tasks. Its adoption at the heart of the

Japanese Fifth Generation Computer Systems project has led to an increase in the volume of research and development work worldwide.

Clocksin, W.F. and Mellish, C.S., *Programming in Prolog*, Springer-Verlag, Berlin, 1994 (Fourth, revised and extended edition).

Bratko, I., *Prolog Programming for Artificial Intelligence*, Addison-Wesley, Wokingham, 1990 (second edition).

Sterling, L. and Shapiro, E.Y., *The Art of Prolog*, MIT Press, Cambridge, MA, 1986.

Paul Brna

217 Proof Editors

While automated Theorem Proving[276] aims to show automatically that a statement is derivable in a formal system, a proof editor is designed so that the user can construct derivations interactively, with the implementation taking care to ensure that the resultant derivation is correct. Proof editors are employed because theorem proving in most logics is difficult or impossible to automate fully.

In practice, systems not only allow the user to invoke rules of inference individually but also allow for some automation, for example by incorporating decision procedures for some classes of problems, or by allowing chains of inference applications to be applied automatically. A procedure which applies an appropriate chain of inference rules to a given problem is often called a tactic.

A proof editor, apart from maintaining the logical correctness of proofs, should ideally aid in operations such as instantiation of variables, recall of earlier proof states, recording of dependencies between proofs, and manipulation of partially specified proofs. The user may be aided by indication of possible 'next steps' in the construction.

Most proof editors are tailored to a particular logic, but there has recently been interest in proof editors that can be customised over a large range of logics.

Huet, G. and Plotkin, G.D., *Logical Frameworks*, Cambridge University Press, Cambridge, 1991.

Alan Smaill

218 Propagation in Cellular Arrays

One particular facet of the use of parallel cellular arrays for image processing is that the physical interconnection structure of the array can reflect the supposed connectivity of images. This approach is most useful when the array allows 'global propagation'. That is propagation of signals from pixel groups across arbitrary distances, with each intermediate pixel transforming the signal according to its local data. For example global propagation allows a CLIP series machine to extract a labelled component of a binary image of arbitrary complexity in one machine cycle. See also Relaxation Labelling[233] and Cellular Arrays[36].

Duff, M.J.B. and Fountain, T.J. (eds.), *Cellular Logic Image Processing*, Academic Press, London, 1986, pp. 42–68.

Dave Reynolds

219 Property Lists

A technique whereby a list of name-value pairs is associated with an object (such as a node of a Semantic Net[244]). Each name represents a certain attribute of the object and the corresponding value records the value of that attribute assigned to the object. Such a facility for associating property lists with atoms is provided as a primitive operation in Lisp[150] type languages.

Allen, J.R., *Anatomy of LISP*, McGraw-Hill, New York, 1978.

Lincoln Wallen

220 Protocol Analysis

A technique for extracting the procedures used by a human problem solver from a record of selected aspects of his/her problem solving behaviour. The record may be of verbal behaviour (e.g., a 'think-aloud protocol') or of non-verbal behaviour (e.g., eye-movements, or the sequence and timing of moves in a board game). The analysis begins with the building of a problem behaviour graph which serves as a 'rational reconstruction' of the solution steps. Regularities are then sought and extracted, possibly in the form of a Production Rule System[213]. Protocol analysis remains an art learned by apprenticeship.

Newell, A., *On the analysis of human problem solving protocols*, Thinking: Readings in Cognitive Science (Johnson-Laird, P.C. and Wason, P.N., eds.), Cambridge University Press, Cambridge, 1977, pp. 46–61.

Ericsson, K.A. and Simon, H.A., *Protocol Analysis: Verbal Reports as Data*, MIT Press, Cambridge, MA, 1984.

Richard Young

221 Pyramids

(*Resolution Cones*)

Strictly, a pyramid/resolution cone is a set of arrays used to store a digitised image at different resolutions, the highest resolution forming the base of the pyramid. Normally, there is a uniform relationship between adjacent levels of the pyramid, for example, a pixel intensity at a particular resolution is often the average value of the pixel intensities in the corresponding 2 x 2 pixel block at the next highest resolution. More generally, the concept of a regular hierarchical data structure has been extended to include processing elements within the structure, then known as a processing cone. Originally, pyramids were used to explicate the relationship between the different image resolutions in a computer vision system using planning. For instance, the boundaries found at a low resolution might be used to refine the Edge Detection[85]/ Boundary Detection[25] process at higher resolutions. This can significantly improve the efficiency of the overall edge detection process. More recently, it has been realised that image boundaries occur over a wide range of scales from sharp step-like edges to fuzzy, blurred edges. Some results have been stated on using the relationships between corresponding edges occurring at different resolutions to make assertions about the scene boundary producing them. Since these regular hierarchical structures are inappropriate for non-spatial symbolic computation their use is normally restricted to the early processing stage of a vision system.

Tanimoto, S.L., *Computer Vision Systems* (Hanson, A.H. and Riseman, E.M., eds.), Academic Press, New York, 1978.

Bob Beattie

222 Quad Trees

A hierarchical image representation similar to Pyramids[221], they have nodes that correspond to the cells of a pyramid and each non-terminal node has four children, but unlike a pyramid, a quad tree may be pruned so as to be unbalanced. For example, when all nodes in a subtree have the same gray value, the subtree may be represented by its root without loss of information, so that significant storage savings can be obtained for many images. More significantly, quad trees allow some operations to be performed efficiently by recursive procedures.

Cohen, P.R. and Feigenbaum, E.A. (eds.), *The Handbook of Artificial Intelligence*, vol. 3, William Kaufmann, Los Altos, CA, 1982.

Luis Jenkins

223 Qualitative Reasoning

Qualitative reasoning involves techniques for reasoning with qualitative, rather than numerical values over time. The simplest systems just use the values 'negative', 'zero' and 'positive' and, given a set of qualitative relations, predict all possible qualitative behaviours. It should be contrasted with Naive Physics[175], which seeks to model human reasoning about the physical world. More complicated systems, like those using envisioning, predict the qualitative behaviour of devices, and provide a theory of causality that can be used to produce causal explanations acceptable to humans.

Qualitative reasoning applies to many areas where exact numerical reasoning is inappropriate or impossible. These include the diagnosis, control, design and explanation of physical or biological systems.

Weld, D.S. and de Kleer, J. (eds.), *Readings in Qualitative Reasoning about Physical Systems*, Morgan Kaufmann, San Mateo, CA, 1990.

Cohn, A.G., *Approaches to qualitative reasoning*, Artificial Intelligence Review **3** (1989), 177–232.

Alison Cawsey, Kevin Poulter

224 Range Finding

There are several methods for obtaining image depth information by range finding. In light striping (a form of Structured Lighting[265]), a single plane of light is projected onto a scene. An imaging system is used to detect this reflected light. If the equation of the light plane and the parameters of the imaging system are known then the position in 3-D space of each point along the line can be determined. A spot range can be determined by emitting a short burst of light (usually with a laser) and measuring the 'time of flight' of the reflected pulse. An alternative approach is to modulate the light and obtain an estimate of the range from the phase change in the returned signal. Instead of light, ultrasound or 'sonar' can also be used. Range estimates are normally achieved by 'time of flight' but ultrasound has the advantage that, unlike light, it is only partially reflected from boundaries enabling range information to be determined through a column of material. The disadvantage is that most surfaces behave like 'sonar mirrors' so that interpreting where you are is like walking in a room full of mirrors.

Ballard, D.H. and Brown, C.M., *Computer Vision*, Prentice-Hall, Englewood Cliffs, NJ, 1982, pp. 52–56.

H.W. Hughes, Ashley Lotto

225 Recursion

(*Recursive Function*)

Recursion theory has been developed by logicians (mainly K. Gödel and S.C. Kleene) in the 1930's as a way to formalise metamathematics. Recursive functions can be given by definitions involving reference to the function being defined. For instance:

$$f(n) = \begin{cases} 1 & \text{if } n=0 \\ n * f(n-1) & \text{otherwise} \end{cases}$$

where n is a natural number, is the recursive definition of the factorial.

Since recursive functions are all and only the functions computable by idealised computers (Turing machines), recursion theory has been extended to domains different from the natural numbers (e.g., lists) and used by computer scientists as a powerful tool to reason about programs. Applications include the theory of computation, complexity of programs, program synthesis and program verification.

Péter, R., *Recursive Functions in Computer Theory*, Ellis Horwood, Chichester, 1981.
Kleene, S., *Introduction to Metamathematics*, Van Nostrand, New York, 1952.

Fausto Giunchiglia

226 Recursion Analysis

This is a technique for constructing an appropriate rule of mathematical induction to prove a theorem. The standard form of mathematical induction is:

$$\frac{P(0) \; \forall n P(n) \rightarrow P(n+1)}{\forall n P(n)},$$

but there are infinitely many such rules: for instance, there is one for each well ordering of each recursively defined data-structure. Thus applying induction presents an infinite branching point in a Search Space[260].

Recursion analysis uses the definitions of the Recursive Functions[225] in a theorem as a clue to the form of induction to be used to prove it. Each recursive definition suggests a dual induction rule. These suggestions are merged together to form an induction rule that subsumes as many of these recursive definitions as possible. This choice of induction rule improves the chances that the definitions of the recursive functions in the theorem can be used during the subsequent proofs of the base and step cases of the induction.

Recursion analysis was invented by Boyer and Moore and implemented in their Nqthm inductive theorem prover. The best explanation can be found in the analysis by Stevens.

Note: mathematical induction is a rule of deduction and is not to be confused with inductive learning, which is a rule of conjecture.

Stevens, A., *A rational reconstruction of Boyer and Moore's technique for constructing induction formulas*, Proceedings of ECAI-88, 1988, pp. 565–570.

Alan Bundy

227 Reflectance Map

A reflectance map relates image intensities to surface orientation for given reflectance function, viewpoint and illumination direction. It is usually possible to construct a reflectance map for a surface even under complicated illumination conditions though it may be necessary to measure the intensities empirically as the relation is frequently too complex to be modelled analytically. See also Intrinsic Images[134].

Horn, B.K.P., *Understanding image intensities*, Artificial Intelligence **8** (1977), 201–231.

T.P. Pridmore, S.R. Pollard, S.T. Stenton

228 Reflection

(*Self-reference*)

It has been recognised by philosophers since classical times that self-reference (e.g., 'this sentence is false') raises difficult issues in the philosophy of language, but only in this century has real progress on analysing the problems been made by, e.g., Russell, Tarski and Gödel. However, while self-reference can be difficult to treat, it is also clear that it offers powerful tools, and is anyway sometimes unavoidable, for example in the design of operating systems or ambitious knowledge representation and processing systems.

Research in mathematical logic has concentrated on Gödel's incompleteness results, which show that that given a suitable true theory T of, e.g., arithmetic, statements (known as *reflection principles*) such as 'T is consistent' are unprovable in T even though they are recognisably true. However given a theory T and reflection principle R, there is nothing to stop us defining a new theory $T + R$ which must also be a true theory of arithmetic which is properly stronger than T (since it can prove every theorem of T, as well as the new addition R). This procedure can even be iterated. A good exposition of this work, and further references, can be found in Girard.

On the other hand, work in philosophical logic and theoretical AI has, for the most part, started from Tarski's 'no truth definition' theorem, which says that it is is not possible to define in a true theory T a predicate *true* such that *true*('A') $\iff$ A is theorem of T for every A, where 'A' is a string with intended meaning A. Effort has concentrated on finding special circumstances where such a *true* might, after all, be defineable, or finding restricted classes of formulae for which it might be defineable, and then investigating the properties of such systems. The results have often exhibited properties similar to modal logics[165]. Bartlett provides an encyclopaedic collection of readings surveying this work.

Finally, there is the more pragmatic area of 'procedural' reflection, initiated by a doctoral thesis by Brian Smith. This investigates the properties of real computer systems which are able to make more or less radical modifications to their own running interpreters. The two commonest foundations for this work are versions of Lisp[150] (allowing programs explicit access to an abstract 'denotational semantics' style representation of the interpreter state) and Smalltalk[185] (programs are able to patch the interpreter binary on the fly). Maes and Nardi provide a useful, though now aging, collection of papers on procedural reflection for programming systems and also for more general knowledge based systems.

Bartlett, S. (ed.), *Reflexivity*, North-Holland, Amsterdam, 1992.

Girard, J.-Y., *Proof Theory and Logical Complexity*, vol. 1, Bibliopolis, Naples, 1987.

Maes, P. and Nardi, D. (eds.), *Metalevel Architecture and Reflection*, North-Holland, Amsterdam, 1988.

Seán Matthews

229 Refutation Proof

(*Proof by Contradiction, Reductio ad Absurdum*)

A method of proof in which the conjecture is negated and a contradiction deduced thus proving the conjecture to be true. This is the method of proof utilised by most Resolution[234] theorem provers.

Nilsson, N.J., *Principles of Artificial Intelligence*, Tioga Pub. Co., Palo Alto, CA, 1980, pp. 161–191, and Springer-Verlag, Berlin, 1982.

Dave Plummer

230 Region Finding

(*Region Growing*)

The basic idea of region finding is to produce a segmentation of the image in which the regions (connected sets of pixels) found have a useful correspondence to projections of scene entities such as objects or surfaces. As such it is the (currently unpopular) dual of Edge Detection[85]/Boundary Detection[25]. There are two main approaches:

- Start with the maximum number of regions (e.g., make every pixel a region) and merge adjacent regions based on some measure of similarity until a satisfactory segmentation has been achieved.
- Start with a few (possibly one) large regions and recursively split them into smaller regions based on some measure of dissimilarity until a satisfactory segmentation has been achieved.

Measures of similarity/dissimilarity have ranged from simple average intensities to complex methods incorporating semantics.

Ballard, D.H. and Brown, C.M., *Computer Vision*, Prentice-Hall, Englewood Cliffs, NJ, 1982.

Zucker, S.W., *Region growing: childhood and adolescence*, Computer Graphics and Image Processing **5** (1976), 382–399.

Bob Beattie

231 Regularisation

A method, for determining a solution to an ill-posed mathematical problem, by constraining the solution to be, in some sense, regular or smooth. An ill-posed problem is one for which there does not exist a unique stable solution. The best known regularisation method is that of minimising (with respect to the solution sought) a functional which consists of two terms, the first representing the fidelity of the solution to the data, and the second its smoothness using a constraint measure chosen from physical considerations of the original problem.

Poggio, T., Torre, V. and Koch, C., *Computational vision and regularisation theory*, Nature **317** (1985), 314–319.

R.M. Cameron-Jones

232 Relational Database Query Formulation

A technique for specifying queries and other interactions against the relational and entity-relationship data models. Database semantics are initially presented as a hierarchy of functional areas, leading to entity-relationship/relational description. A naive-user mode provides system initiated dialogue with natural language statements of queries (i.e., paraphrases of relational calculus expressions) displayed for user validation. Self-teaching system, with simple 'user controlled' inferencing.

Longstaff, J., *Controlled inference and instruction techniques for DBMS query languages*, P111 Proceedings of 1982 ACM SIGMOD Conference, Association for Computing Machinery (1982).

Jim Longstaff

233 Relaxation Labelling

Relaxation labelling is a technique for assigning globally consistent labels or values to nodes in a network subject to local constraints, by iteratively propagating the effects of Constraints[52] through the net. It has its mathematical origins as a technique in numerical analysis for the solution of difference equations and recent developments have shown it to be related to various optimisation techniques e.g., linear programming. The first significant application of relaxation labelling to a vision problem was Waltz's filtering algorithm in the blocks world line labelling domain. Consider the problem of assigning labels to objects to satisfy certain consistency requirements. Unlike a tree representation where each context is explicitly a path, the space may be represented as a graph in which each node carries a set of possible labels. The task is to find a single labelling for each node that satisfies the set of constraints. In general, after an initialisation stage in which each node has been assigned a list of labels and their associated confidence measures, the labels and confidences of neighbouring nodes are compared and, guided by the requirement to minimise local inconsistency (often a smoothness constraint), labels are deleted or their confidences adjusted. This process of comparison and adjustment iterates until it converges to some criterion of global consistency. Because both the assignment and updating processes can be done independently at each node, the computation is inherently parallel.

Apart from Waltz's classical application, relaxation labelling has been used in the computation of Optical Flow[190], the recovery of surface orientation from Shading Information[249], and the recovery of the orientation structure of images, Stereopsis[261] and Structure from Motion[266]. The convergence properties of some relaxation operators are not always transparent. The most successful and scientifically useful applications have been when the theoretical analysis of a vision problem reveals a mathematical structure that can be directly exploited in the design of the relaxation algorithm rather than when a 'general operator' has been fitted in an ad hoc fashion to a vision labelling task. See also propagation in Cellular Arrays[36].

Davis, L.S. and Rosenfeld, A., *Cooperating processes in low level*, Artificial Intelligence **17** (1981), 246–265.

Ballard, D.H. and Brown, C.M., *Computer Vision*, Prentice-Hall, Englewood Cliffs, NJ, 1982, pp. 408–430.

Jon Mayhew

234 Resolution

A rule of inference of Predicate Calculus[208] used to deduce a new formula from two old ones. It has been used extensively in automatic Theorem Proving[276], because it is an efficient alternative to traditional rules of inference in mathematical logic. All the three formulae involved must be in Clausal Form[41]. If C and D are clauses and the $\mathrm{P_i}$ and $\mathrm{Q_j}$ are atomic formulae then the rule is

$$\frac{\mathrm{C} \vee \mathrm{P}_1 \vee \ldots \vee \mathrm{P}_m \qquad\qquad \mathrm{D} \vee \sim \mathrm{Q}_1 \vee \ldots \vee \sim \mathrm{Q}_m}{(\mathrm{C} \vee \mathrm{D})\theta},$$

where θ is the most general unifier of all the $\mathrm{P_i}$ and $\mathrm{Q_j}$, and is obtained by Unification[286]. Resolution is the execution mechanism of Prolog[216].

Chang, C. and Lee, R.C., *Symbolic Logic and Mechanical Theorem Proving*, Academic Press, New York, 1973.

Alan Bundy

235 Rhetorical Structure Theory

(RST)

Rhetorical Structure Theory (RST) aims to provide an account of the phenomenon of *text coherence*. While many linguistic theories are concerned primarily with the structure of single sentences, RST explores the ways in which clauses and sentences can combine to form larger units of text, such as paragraphs, reports, or articles.

The primary assumption underlying the theory is that a coherent text contains more information than is provided by its clauses and sentences individually. The additional information emerges from the way these units are related together. For instance, two clauses can be related as cause and effect, or as elements in a temporal sequence, or as premise and conclusion. RST proposes a set of some 23 *rhetorical relations*, such as *volitional cause*, *contrast* and *justify*, which it claims are sufficient to analyse 'the vast majority' of coherent texts. These relations are defined *functionally*; that is to say, in terms of the effects the writer intends to achieve on the reader by the juxtaposition of the two related units. While relations can sometimes be signalled linguistically (for instance by sentence or clause connectives), they are defined without any reference to surface linguistic forms.

The units linked by relations are known as *text spans*. Spans can be of any size, from single clauses upwards. Relations map onto texts in the first instance via structures called *schemas*, which group a small number of adjacent text spans in various different ways. The spans grouped by a schema are treated as a new composite span, which can then itself be related to other spans by other rhetorical relations. In this way a hierarchical pattern of schemas is built up, known as a *rhetorical structure tree*. A coherent text is characterised as one which can be described by such a tree.

RST has recently found widespread application in the field of natural language generation. The hierarchical/compositional nature of rhetorical relations, as well as their definition in terms of the writer's goals, make them well-suited for use as *planning operators*, in the tradition of classical planning systems like STRIPS. Current debate about RST centres around the choice of the set of relations to be implemented and on how they should be defined, as well as on the question of what other mechanisms (if any) are necessary in addition to relations for a complete account of text coherence.

Mann, W.C., Matthiessen, C.M. and Thompson, S.A., *Rhetorical Structure Theory and Text Analysis* (1989), Information Sciences Institute, Marina del Rey, CA (RR/89/242).

Hovy, E., *Automated discourse generation using discourse structure relations*, Artificial Intelligence **63** (1993), 341–385.

Moore, J. and Paris, C., *Planning text for advisory dialogues: capturing intentional and rhetorical information*, Computational Linguistics **19** (1993), 651–694.

Alistair Knott

236 Rewrite Rules

(*Condition-action Pairs, Demodulants*)

Rewrite rules are sets of ordered pairs of expressions ⟨lhs,rhs⟩ usually depicted as (lhs $\Longrightarrow$ rhs). There is usually a similarity relation between the 'lhs' and the 'rhs' such as equality, inequality or double implication. Rewrite rules, as the pairs are called, together with the rewriting rule of inference allow one expression to be 'rewritten' into another. A subexpression of the initial expression is matched with the 'lhs' of the rewrite rule yielding a substitution. The resulting expression is the expression obtained by replacing the distinguished subexpression with the 'rhs' of the rewrite rule after applying the substitution.

The matching process may be full Unification[286] or, more usually, a restricted form of Pattern Matching[197] where only the variables in the rewrite rule may be instantiated. Examples of the use of rewrite rules are the restricted Paramodulation[192] inferences called demodulation performed in theorem provers, or programming with abstract data types introduced by a series of equations. Some powerful theoretical results have been obtained for rewriting systems.

Huet, G. and Oppen, D.C., *Equations and rewrite rules: a survey*, Formal Language Theory: Perceptives and Open Problems (Book, R.V., ed.), Academic Press, New York, 1980.

Lincoln Wallen

237 Robot Dynamics

Robot dynamics addresses the problems of calculating the acceleration of a robot from a given set of forces (forward dynamics), and calculating the forces required to produce a given acceleration (inverse dynamics). Forward dynamics is used for simulation, inverse dynamics for control. A robot is considered to be a system of rigid bodies, or links, connected together by joints. The laws of motion for rigid bodies are used to find the equations of motion for the robot. The two main approaches are via Newton's and Euler's equations, and via Lagrange's equation. The tricky part is to calculate the answers efficiently. This is done by expressing the equations of motion in terms of recurrence relations between the motion of one link and that of one of its neighbours in such a way that the answers may be calculated recursively.

Hollerbach, J.M., *A recursive Lagrangian formulation of manipulator dynamics and a comparative study of dynamics formulation complexity*, IEEE Trans. Systems, Man & Cybernetics **SMC-10** (1980), no. 11, 730–736, also in Robot Motion: Planning and Control (Brady, M., et al., eds.), MIT Press, Cambridge, MA, 1982, pp. 73–87.

Walker, M.W. and Orin, D.E., *Efficient dynamic computer simulation of robotic mechanisms*, Trans. ASME, ser. G, Jnl. Dyn. Sys. Measurement & Control **104** (1982), 205–211, also in Robot Motion: Planning and Control (Brady, M., et al., eds.), MIT Press, Cambridge, MA, 1982, pp. 107–125.

Paul, R.P., *Robot Manipulators*, MIT Press, Cambridge, MA, 1981.

Roy Featherstone

238 Scale-Space Representation

The representation of an image quantity (e.g., zero-crossings) over a range (which may in theory vary continuously) of scales at which the image is perceived. The 'scale' concerned is generally the width parameter of a Gaussian function with which the image is convolved; thus at small scales the image detail is faithfully represented and at large scales the detail is blurred as the result tends to the image mean.

Witkin, A.P., *Scale-space filtering*, Proceedings of IJCAI-83, 1983, pp. 1019–1022.

R.M. Cameron-Jones

239 Script

A structure for the large scale organisation of knowledge, adopted by Schank primarily as a support for natural language understanding, and related to Conceptual Dependency[45] as the primary form of knowledge representation. Scripts define the normal character and sequence of events in, for example, a restaurant visit as a temporal incident. They can thus be used to assign an order to language data that do not give temporal information explicitly, and may also be used to indicate underlying causal relationships. The need for explicit inference to determine temporal or causal relations between data instances is therefore reduced. In Schank's view the event orientation of scripts distinguishes them from other Frames[98], but they share other properties of frames, e.g., defaults and attached procedures, and forms of set organisation, and present similar problems of definition and use. Scripts have been applied to a wide variety of language processing tasks (e.g., MOPS), chiefly by the Yale group. In general, usage of the term is less variable than that of frame, but it is still applied with a good deal of freedom.

Schank, R. and Abelson, R., *Scripts, Plans, Goals and Understanding: An Inquiry into Human Knowledge Structures*, Lawrence Erlbaum Associates, Hillsdale, NJ, 1977.

Collins, A. and Smith, E.E. (eds.), *Readings of Cognitive Science: A Perspective from Psychology and Artificial Intelligence*, Morgan Kaufmann, San Mateo, CA, 1988 (Chapters 1–3 and pp. 190–223).

Karen Sparck Jones

240 Self-organising Feature Maps

Self-organising feature maps were introduced by Kohonen in 1982. They perform a dimensionality reduction, with a multi-dimensional input being projected onto a two-dimensional feature map. A key notion here is that of conservation of topology—points that are close together in input space are mapped in such a way that they remain close in the two-dimensional feature space. This notion of dimensionality reduction is obviously inapplicable to uniformly distributed input data; however it transpires that many real-world problems may be mapped in such a way without a catastrophic loss of information. An example of the use of this algorithm is in speech recognition, whereby high-dimensional speech data is mapped onto a 2-dimensional 'phonotopic map'.

Kohonen, T., *Self-organisation and Associative Memory*, (Springer Series in Information Sciences, vol. 8), Springer-Verlag, Berlin, 1989 (3rd edition).

Steve Renals

241 Semantic Checking

(*Use of Models*)

A technique for pruning a Search Space[260] of a logical inference mechanism, e.g., Resolution[234] or Natural Deduction[176]. One or more models are given of the axioms and hypotheses of a problem. If a model is not a counterexample then the goal is also true in that model. All subgoals false in any model are pruned from the search space. This technique preserves completeness if the problem consists only of Horn clauses.

Gelernter, H., *Realization of a geometry theorem proving machine*, Computers and Thought (Feigenbaum, E.A. and Feldman, J., eds.), McGraw-Hill, New York, 1963, pp. 134–152.

Bundy, A., *The Computer Modelling of Mathematical Reasoning*, Academic Press, London, 1983, pp. 169–189, also second edition, 1986.

Alan Bundy

242 Semantic Grammar

Semantic grammar is contrasted with conventional grammars as it relies predominantly on semantic rather than syntactic categories, e.g.,

MESSAGE → PATIENT-TYPE, HAVE, DISEASE-TYPE.

In some cases the semantic categories and structures are merely cosmetic relabellings of conventional syntactic categories and structures, but more thorough semantic grammars are widely used, though even these typically contain some mixture of syntactic elements. Semantic grammars have been found especially effective for language processing in limited domain contexts, e.g., processing medical records, interpreting database queries, where syntactic parsing is unnecessarily costly, but general-purpose semantic grammars have also been proposed. There is a connection between these grammars and Semantic Primitives[245] and semantic Case Frames[33]. However systems making heavy use of general-purpose semantics are not conventionally described as relying on semantic grammars.

Brown, J.S. and Burton, R., *Multiple representations of knowledge for tutorial reasoning*, Representation and Understanding (Bobrow, D. and Collins, A., eds.), Academic Press, New York, 1975, pp. 311–349.

Karen Sparck Jones

243 Semantic Head-Driven Generation

(*Head-driven Bottom-up Generation*)

Semantic Head-Driven Generation is a technique for generating a sentence (or a phrase) from its corresponding semantic representation. The algorithm traverses the derivation tree of the sentence in a semantic-head-first fashion. Starting from a node (*root*), which contains the input semantics and some syntactic information (like the category of the phrase to be generated), the algorithm selects a grammar rule in which the semantics of the mother node is identical to the semantics of the root. If the selected grammar rule passes the semantics unchanged from the mother node to a daughter node (*chain rule*), the daughter node becomes the new root and the algorithm is applied recursively (this is the top-down prediction step). If a grammar rule is reached which does not pass the semantics unchanged from the mother node to a daughter node (*a non-chain rule*) then the grammar rule is expanded and the algorithm is recursively applied to the list of the daughters (the mother node of the non-chain rule is called the *pivot*). The top-down base case occurs when the non-chain rule has no non-terminal daughters; that is, it introduces lexical material only. When all daughters of a non-chain rule have been generated, the algorithm attempts to 'connect' the pivot to the root. This is done in a bottom-up fashion using chain rules. Larger and larger trees are built which contain the tree dominated by the pivot. This proceeds until the top node of the new tree is compatible with the initial root.

The Semantic Head-Driven Generation algorithm is an illustrative instance of a sophisticated combination of top-down prediction and bottom-up structure building. It properly handles left-recursive rules and semantically non-monotonic grammars which have been problematic for previous purely top-down and bottom-up generation strategies respectively. The overall processing is directed by the input logical form and the information in the lexicon which leads to a goal-directed generation with good performance results.

Shieber, S.M., van Noord, G., Moore,R.C. and Pereira, F.C.N., *A semantic head-driven generation algorithm for unification-based formalisms*, Computational Linguistics **16:1** (1990), 30–42.

van Noord, G., *An overview of head-driven bottom-up generation*, Current Research in Natural Language Generation (Dale, R., Mellish, C. and Zock, M., eds.), Academic Press, London, 1990, pp. 141–165 (Chapter 6).

Nicolas Nicolov

244 Semantic Networks

(*Semantic Nets*)

Semantic nets are a principle for the large scale organisation of knowledge emphasising the multiple associations of individual concepts. Concepts, objects, entities, etc. are represented as nodes in a linked graph, and relationships between these are represented as labelled arcs. The range of possible network structure types is very wide (see Findler). Semantic nets should properly be based on definitions of the net structure, i.e., the syntax and semantics of nodes and links and of configurations of these, and of net operations, i.e., the syntax and semantics of node-node transitions, but too frequently are not. Nets have been found an attractive descriptive device, but genuine algorithmic exploitation of nets based, e.g., on the general idea of Marker-passing[156] for selective reading or writing at nodes, is comparatively rare (formal graph theory is rarely seriously invoked in Artificial Intelligence). The emphasis on concept association introduces difficulties in representing any partitioning or grouping of net elements, for example to represent quantified propositions, clusters of similar entities, etc. (but see Partitioned Net[194]), and network searching conspicuously manifests the general AI problem of the combinatorial explosion.

Findler, N.V. (ed.), *Associative Networks: Representation and the Use of Knowledge by Computers*, Academic Press, New York, 1979.

Sowa, J. (ed.), *Principles of Semantic Networks: Explorations in the Representation of Knowledge*, Morgan Kaufmann, San Mateo, CA, 1991.

Lehmann, F. (ed.), *Semantic Networks in Artificial Intelligence*, Pergamon Press, Oxford, 1992.

Karen Sparck Jones

245 Semantic Primitives

The general concepts underlying words, used for the determination and representation of textual or propositional meaning, e.g., MOVE underlies 'walk' and 'run', THING 'vase' and 'book'. Semantic primitives are used to define selection criteria for sense identification, and to define key properties, e.g., of Case Frame[33] role fillers. Primitives, which should in principle be drawn from a closed set, may be shallow or deep, i.e., more or less fine-grained, domain-independent or domain-dependent, categorial or relational, etc. The characterisations of word (senses) may be by single primitives, primitive sets, or structured formulae, and the primitive characterisations of sentences may be more or less elaborately structured. The primitive names may be regarded as elements of a distinct meaning representation language, or as selected elements of the language under description. Though primitives universally figure in some form or other in language understanding systems, they are more frequently adopted ad hoc than systematically motivated. Wilks and Schank are exceptions here.

Charniak, E. and Wilks, Y., *Computational Semantics: An Introduction to Artificial Intelligence and Natural Language Comprehension*, North-Holland, Amsterdam, 1976.

Sowa, J., *Conceptual Structures: Information Processing in Mind and Machine*, Addison-Wesley, Reading, MA, 1984, pp. 103–115.

Karen Sparck Jones

246 Semantic Syntax

(*SeSyn*)

Semantic Syntax (SESYN) is an integrated theory of natural language syntax and semantics. It is a direct continuation of Generative Semantics. SESYN is a rule system that establishes a (bidirectional) mapping between the meaning representation of sentences (called semantic analysis structures) and their surface realisation. The semantic analysis structures are higher-order predicate calculus trees and contain the lexical items for open class words. The surface structures are based on an orthodox version of Transformational Grammar.

The rule system works as follows: first a semantic analysis structure is constructed using a set of formation rules. The formation rules are a grammar for the semantic analysis structures. Then, the semantic analysis structure is transformed into a surface structure using cyclic and post-cyclic transformational rules. The cyclic rules apply in a bottom-up way and are mostly lexicon-driven: predicates are lexically marked for the cyclic rules they induce. The post-cyclic rules are largely structure-driven and apply in linear order as defined by the grammar. There is a limited number of highly constrained transformation rules. In SESYN there is no surface syntax grammar. Instead, a grammar for the semantic structure is used, coupled with a transformational component—SESYN is a procedural framework. SESYN is formally precise and achieves a high degree of empirical success. Exact rule systems are available for a number of languages (English, French, Dutch, German) and more are being developed.

Seuren, P.A.M., *Semantic Syntax*, Blackwell, Oxford, 1996.

Nicolas Nicolov

247 Sensory Feedback and Compliance

A common problem in robotics is to bring a tool to a certain spatial relationship to a workpiece, or to bring two or more components together into a certain spatial relationship. However, owing to accumulation of dimensional variation and tolerances that cannot be minimised at their source, errors can occur, making it unexpectedly difficult or impossible to solve this problem by dead reckoning. Thus, in many cases, a robot must be able to detect that components are poorly positioned, and it must have the ability to make small incremental changes in position to minimise the error. 'Sensing', or the measurement of physical properties (optical, acoustic, tactile, etc.) can be used actively to detect displacements from the correct position; the resulting error signal is fed back to steer the robot in the direction to minimise the error. This is known as closed-loop control, which is formalised by control theory. Compliance is a passive control technique, where springs and other mechanical devices are used to produce corrective displacements in reaction to forces in the tool.

Simons, G.L., *Robots in Industry*, NCC Publications, 1980.
Nevins, J.L. and Whitney, D.E., *Computer controlled assembly*, Scientific American **238** (1978), no. 2, 62–74.
Dorf, R.C., *Modern Control Systems*, Addison-Wesley, Reading, MA, 1980.

W.F. Clocksin

248 Sequent Calculus

A method of presenting a logic invented by Gentzen in which statements are made in the form:

$$A_1, A_2, \ldots A_n \vdash B_1, B_2, \ldots B_m$$

where the sequent sign '$\vdash$' indicates that given all the propositions A_i one of the propositions B_j holds. A logic is characterised by giving basic sequents as axioms and then allowing further sequents to be derived using rules of inference, typically associated with the introduction and elimination of the logical symbols of the language on either side of the arrow; for example, the rule for introducing a disjunction is:

$$\frac{\Gamma \vdash \Theta, P, Q}{\Gamma \vdash \Theta,\ P \vee Q},$$

where Γ and Θ represent an arbitrary number of propositions. One of the advantages of sequent calculus over, say, Natural Deduction[176] is the ease with which proofs can be manipulated (e.g., Gentzen's work on proof theory).

Gallier, J.H., *Logic for Computer Science: Foundations of Automatic Theorem Proving*, Harper & Row, New York, 1986.

Alan Smaill

249 Shape from Shading

The process of extracting three-dimensional shape information from smooth gradations of reflected light intensity. It has been shown by Horn that if certain assumptions are made concerning the reflectance function and illumination of a surface it is possible to formulate and solve equations relating surface shape to the measured intensities in an image of the surface. See Intrinsic Images[134].

Horn, R.K.P., *The Psychology of Computer Vision* (Winston, P.H., ed.), McGraw-Hill, New York, 1975.

T.P. Pridmore, S.R. Pollard, S.T. Stenton

250 Supervised Learning

In supervised learning the learner is presented with pairs of input-output patterns. Its task is to infer a more compact representation of the mapping between these sets. Supervised learning methods are usually incremental (with the notable exception of ID3[40]). Each pair of input-output patterns is presented in turn, and the whole set of patterns, called the training set, may be presented several times before learning is complete. Compare this technique with Unsupervised Learning[289].

If a learning procedure is incremental it may get stuck, by converging to a fundamentally incorrect approximation to the true function. This phenomenon is sometimes described in terms of becoming stuck in a local minimum within the space of possible approximations to the function. Some of the procedures use gradient descent on an error surface (e.g., the delta rule of back-propagation). Others start with a space of possible descriptions for a concept and reduce that space according to the training pairs presented (Focussing[92], AQ11). ID3 uses information theory to build the most compact representation of the mapping it can in the form of a discrimination net.

Rumelhart, D.E., Hinton, G.E. and Williams, R.J., *Parallel Distributed Processing: Explorations in the Microstructure of Cognition, Volume 1* (Rumelhart, D.E., McClelland, J.L. and The PDP Research Group, eds.), MIT Press, Cambridge, MA, 1986, pp. 318–362.

Mitchell, T., *Generalisation as search*, Artificial Intelligence **18** (1982).

Dietterich, T., London, B., Clarkson, K. and Dromey, G., *The Handbook of Artificial Intelligence*, vol. 3 (Cohen, P.R. and Feigenbaum, E.A., eds.), William Kaufmann, Los Altos, CA, 1982.

Jeremy Wyatt

251 Shape from Texture

Shape from texture is the process by which the three-dimensional structure of a surface is determined from the spatial distribution of surface markings. Projection into the image plane distorts the geometry of surface textures in a manner dependent upon the shape of the underlying surface: as a surface recedes from the viewer its markings appear smaller due to perspective effects, and as a surface is inclined away from the frontal plane its markings appear compressed in the direction of the inclination. By isolating these projective distortions it is possible to recover the shape of a textured surface.

Witkin, A.P., *Recovering surface shape and orientation from texture*, Artificial Intelligence **17** (1981), 17–45.

T.P. Pridmore, S.R. Pollard, S.T. Stenton

252 Simulated Annealing

Simulated annealing is a technique for solving combinatorial optimisation problems by means of the metropolis algorithm. Optimisation requires the use of an objective function (or evaluation function), which represents a quantitative measure of the 'goodness' of a proposed solution. A conventional technique for iterative improvement considers local changes to a proposed solution, accepting just those changes that improve the solution. On realistic problems, this can lead to a local optimum at which all subsequent changes do not improve the solution, even though better solutions are available some distance away. By contrast, in an annealing system, the acceptance of a proposed change depends on a function of both the goodness of the change and a Gaussian noise source. The standard deviation of the noise begins high and is gradually reduced according to a predetermined 'annealing schedule' until no further changes occur. The term 'annealing' is suggested by the formal connection between statistical thermodynamics and combinatorial optimisation. Simulated annealing is being applied in perception and learning research, and in computer-aided design.

Kirkparick, S., Gelatt, C.D. and Vecchi, M.P., *Optimisation by simulated annealing*, Science **220** (1983), 671–680.

W.F. Clocksin, Geoffrey Sampson

253 Situation Calculus

A technique for representing time or different situations in an assertional database. Each relation/property which changes over time is given an extra argument place, which is filled with a situation term. In plan formation this situation argument is usually a nested term representing the sequence of plan steps needed to get to this situation from the initial state. Thus situations label the effect of actions and can represent alternative futures, whereas a numerically measured time can only represent a unique future.

Nilsson, N.J., *Principles of Artificial Intelligence*, Tioga Pub. Co., Palo Alto, CA, 1980, and Springer-Verlag, Berlin, 1982.

Alan Bundy

254 Skeletonisation

This is a transformation of a binary image which locates the central axis ('skeleton' or 'medial-axis') of each object in the image. The general principle is to thin each object in the image to a line structure by repeatedly deleting edge pixels whose demise will not change the connectivity of the image. Algorithms are available which can also deal with three-dimensional binary images.

Tsao, Y.F. and Fu, K.S., *A parallel thinning algorithm for 3-D pictures*, Computer Graphics and Image Processing **17** (1981), 315–331.

Dave Reynolds

255 Skolemisation

A technique borrowed from mathematical logic (and named after the mathematician Skolem), but much used in Automatic Theorem Proving[276], for removing quantifiers from Predicate Calculus[208] formulae. If $A(y)$ is a formula with free variables y, $x_1, \ldots, x_n$, then $\forall y A(y)$ is replaced by $A(y)$, and $\exists y A(y)$ is replaced by $A(f(x_1, ..., x_n))$, where f is a new Skolem function. The technique is usually applied to formulae which have all their quantifiers at the front (prenex normal form), but can be adapted to any formula. It produces a formula which has a model if and only if the original formula does. It is one of the techniques used in turning a formula into Clausal Form[41].

Chang, C.L. and Lee, R.C., *Symbolic Logic and Mechanical Theorem Proving*, Academic Press, New York, 1973.

Alan Bundy

256 Spatial Frequency Channels

Spatial frequency channels are systems sensitive to a limited range of spatial frequencies. In the human visual system these are considered to be a population of cells with similar tuning characteristics specifically sensitive to a restricted range of the Contrast Sensitivity Function[58] envelope.

Wilson, H.R. and Bergen, J.R., *A Fourier mechanism model for spatial vision*, Vision Research **19** (1979), 19–32.

T.P. Pridmore, S.R. Pollard, S.T. Stenton

257 Spelling Correction

An essential requirement for serious natural language processing programs, though not strictly an AI technique.

A simple strategy, based on a letter-by-letter tree-structured dictionary, assumes that errors fall into the four types: (i) extra letter (ii) substituted letter (iii) omitted letter, and (iv) reversed letter pair. Then at any point where a mismatch between input string and dictionary string occurs a new match can be tried by, respectively, (i) advancing the input string one letter (ii) advancing both strings together (iii) advancing the dictionary string, and (iv) advancing first one string and then the other.

This strategy is looking for letter position correspondence between the two strings; weaker strategies look merely for ordinal correspondence, and yet weaker for 'material' correspondence, i.e., just having the same letters. The simple strategy described does not take account of the number of errors per word. Generalising for this requires a string similarity measure. Such measures allow the use of non-literal word representations, e.g., hash coding, and of n-gram rather than single-letter based matching. Spelling correction may also use heuristics exploiting, e.g., distinctive properties of the language ('u' after 'q' in English), those of the input device (optical character reader, human typist), and the choice of strategy may be influenced by the task, e.g., whether a large lexicon is involved, whether a user should be consulted for a proposed correction, etc. In the limit, error detection and correction requires full language understanding.

Pollock, J.J., *Spelling error detection and correction by computer: some notes and a bibliography*, Journal of Documentation **38** (1982), 282–291.

Karen Sparck Jones

258 Standard ML

(ML)

A Functional Language[101] where functions are first-class data objects (i.e., it is higher order); the language is interactive and statically scoped. Standard ML is strongly Typed Language[284] with a polymorphic type system. Abstract data types are supported, together with a type-safe exception mechanism. Standard ML also has a modules facility to support the incremental construction of large programs.

Milner, R., Tofte, M. and Harper, R., *The Definition of Standard ML*, MIT Press, Cambridge, MA, 1990.

Paulson, L., *ML for the Working Programmer*, Cambridge University Press, Cambridge, 1991.

Kevin Mitchell

259 SSS*

This algorithm, introduced by Stockman in 1979, conducts a 'State Space Search' traversing a game tree in a Best-first[116] fashion similar to that of A*[2]. It is based on the notion of solution (or personal) trees. Informally, a solution tree can be formed from any arbitrary Game Tree[10] by pruning the number of branches at each MAX node to one. Such a tree represents a complete strategy for MAX, since it specifies exactly one MAX action for every possible sequence of moves might be made by the opponent.

Given a game tree, G, SSS* searches through the space of partial solution trees, gradually analysing larger and larger subtrees of G, eventually producing a single solution tree with the same root and Minimax[164] value as *G*. SSS* never examines a node that Alpha/Beta Pruning[8] would prune, and may prune some branches that alpha/beta would not. Stockman speculated that SSS* may therefore be a better general algorithm than alpha/beta. However, Rozen and Pearl have shown that the savings in the number of positions that SSS* evaluates relative to alpha/beta is limited and generally not enough to compensate for the increase in other resources (e.g., the storing and sorting of a list of nodes made necessary by the best-first nature of the algorithm). See also DSSS*[80].

Stockman, G.C., *A minimax algorithm better than alpha-beta?*, Artificial Intelligence **12** (1979), 179–196.

Rozen, I. and Pearl, J., *A minimax algorithm better than alpha-beta? Yes and no*, Artificial Intelligence **21** (1983), 199–220.

Shinghal, R., *Formal Concepts in Artificial Intelligence*, Chapman and Hall, London, 1992, pp. 549–569.

Ian Frank

260 State Space

(*Search Space, Problem Space*)

Many problems can be represented as an initial state, a goal state and a set of Operators[187] that define operations to go to new states from a given state. The states that can be reached from the initial state by applying the rules in all possible ways define the state space. The problem is then to reach the goal state from the initial state. By this formulation almost any problem can be reduced to a search problem.

Barr, A. and Feigenbaum, E.A. (eds.), *The Handbook of Artificial Intelligence*, vol. 1, William Kaufmann, Los Altos, CA, 1981, p. 260.

Korf, R.E., *Search: a survey of recent results*, Exploring Artificial Intelligence (Survey Talks from the National Conferences on Artificial Intelligence) (Shrobe, H.E., ed.), Morgan Kaufmann, San Mateo, CA, 1988, pp. 197–237 (Chapter 6).

Maarten van Someren

261 Stereopsis

The process of recovering the 3-D structure of a scene from two different views. The problem has two parts; the measurement of the disparity of corresponding points in the two images, and the interpretation of these disparity measurements to recover the range and orientation of the surfaces in the scene. See Intrinsic Images[134] and Relaxation Labelling[134].

Mayhew, J.E.W., *Stereopsis*, Physical and Biological Processing of Images (Braddick, O.J. and Sleigh, A.C., eds.), Springer-Verlag, New York, 1983, pp. 204–216.

T.P. Pridmore, S.R. Pollard, S.T. Stenton

262 Stochastic Geometry

The representation of parameters associated with geometric primitives as stochastic processes; thus enabling the application of techniques from stochastic filtering theory. It is most commonly used in problems of data fusion, where the (possibly Extended) Kalman Filter is the best known tool as in most application areas of stochastic filtering.

Ayache, N. and Faugeras, O.D., *Building, registrating and fusing noisy visual maps*, International Journal of Robotics Research **7** (1988), no. 6, 45–65.

R.M. Cameron-Jones

263 Stochastic Simulation

(*Gibbs Sampling*)

A method of computing probabilities by counting events in a series of simulation runs. In each run the occurrence of an event is determined by sampling a random distribution, reflecting the state of related events as well as the strength of interaction among them. In Boltzmann machines stochastic simulation is used together with Simulated Annealing[252] to find the state of lowest energy.

Pearl, J., *Probabilistic Reasoning in Intelligent Systems: Networks of Plausible Inference*, Morgan Kaufmann, San Mateo, CA, 1988.

Judea Pearl

264 Structured Induction

Structured induction employs the same top-down problem decomposition as structured programming, combined with bottom-up implementation of the individual subproblems. A given problem is split into relevant attributes, those attributes that are not directly codable are split again. This decomposition process is repeated for each attribute that is not immediately codable until none is left, producing a hierarchical tree of subproblems whose leaf nodes are directly codable attributes. Inductive inference is then used to solve each of the subproblems from the bottom of this hierarchy to the top. Each newly solved subproblem is given a meaningful name which is used in the next level up as a simple coded attribute. This process is continued until there are no more subproblem hierarchy levels to ascend. A top level procedure now exists that when run, calls the lower level subproblems and attributes in an order determined by the inductive procedures applied at each level of the bottom-up implementation. This technique was developed at the Machine Intelligence Research Unit, University of Edinburgh as an aid to the generation of humanly understandable classification rules for use in expert systems.

Shapiro, A. and Niblett, R.B., *Automatic induction of classification rules for a chess end-game*, Advances in Computer Chess, 3, Pergamon Press, Oxford, 1982.

Alen Shapiro

265 Structured Light

The use of the geometry of the illumination actively emitted into the scene to simplify the task of inferring environmental structure. It is often used for Range Finding[224]. The best known example of structured light is illuminating the scene with a (scanning) light-stripe enabling the direct interpretation of breaks in the sensed line as depth discontinuities in the scene. With knowledge of the geometric position of the light source and the relative position of the sensor system, the scene position of the observed stripe can often be calculated.

Popplestone, R.J., Brown, C.M., Ambler, A.P. and Crawford, G.F., *Forming models of plane-and-cylinder faceted bodies from light stripes*, Proceedings of IJCAI-75, 1975, pp. 664–668.

R.M. Cameron-Jones

266 Structure from Motion

The calculation of three-dimensional global structure from the two-dimensional properties of features visible in a time series of images. The features are usually points, but may be lines, etc. Most methods require the correspondence of features between images and rely upon the rigidity assumption, i.e., that the environment is rigid. There are many results on the number of views required for a minimal solution to exist for a given number of points visible in all the images, amongst which the best known (first derived for orthographic projection by Ullman under the rigidity assumption) is that three views of four points are required. In a rigid environment, the method is a form of stereo—the determination of structure from images taken from more than one viewpoint.

Faugeras, O.D., Lustman, F. and Toscani, G., *Motion and structure from motion from point and line matches*, Proceedings of the First International Conference on Computer Vision, 1987, pp. 25–34.

R.M. Cameron-Jones

267 Subgoaling

(*Problem Reduction*)

Many planning systems use a Backwards Search[16] of the space that is defined by the available Operators[187]. The goal is split into subgoals, and the system then recursively tries to satisfy those subgoals. A major problem with this method is that the subgoals may be interdependent. To achieve subgoal G1 it may be necessary to apply an operator that makes it impossible to achieve subgoal G2. See Interactions Between Sub-goals[131].

Nilsson, N.J., *Principles of Artificial Intelligence*, Tioga Pub. Co., Palo Alto, CA, 1980, and Springer-Verlag, Berlin, 1982.

Stefik, M.J., *Planning with constraints (Molgen: part 1) and planning and meta-planning (Molgen: part2)*, Artificial Intelligence **16** (1981), 111–139 and 141–169.

Sussman, G.J., *A Computer Model of Skill Acquisition*, Elsevier, New York, 1975.

Maarten van Someren

268 Superquadrics

Superquadrics are a set of parameterised volumetric shapes used for computer vision applications. They are useful because a wide range of solid shapes can be generated by use of only a few parameters. Surface patches can be defined by the surface of a superquadric volume. Additionally, more complex objects can be defined by combinations of simpler volumes using union, intersection and complement.

Any superquadric may be expressed by the equation:

$$\chi(\eta,\omega) = \begin{pmatrix} a1\cos^{\varepsilon 1}\eta\cos^{\varepsilon 2}\omega \\ a2\cos^{\varepsilon 1}\eta\sin^{\varepsilon 2}\omega \\ a3\sin^{\varepsilon 1}\eta \end{pmatrix},$$

where $-\pi/2 \leq \eta \leq \pi/2$ and $-\pi \leq \omega < \pi$, $a1$, $a2$ and $a3$ are the length, width and breadth respectively. $\varepsilon 1$ and $\varepsilon 2$ are the parameters which specify the shape of the superquadric. $\varepsilon 1$ is the squareness parameter in the north-south direction. $\varepsilon 2$ is the squareness parameter in the east-west direction. The advantage of this representation is that the normal vectors are given by:

$$n(\eta,\omega) = \begin{pmatrix} \frac{1}{a1}\cos^{2-\varepsilon 1}\eta\cos^{2-\varepsilon 2}\omega \\ \frac{1}{a2}\cos^{2-\varepsilon 1}\eta\sin^{2-\varepsilon 2}\omega \\ \frac{1}{a3}\sin^{2-\varepsilon 1}\eta \end{pmatrix},$$

enabling the shape parameters to be recovered via surface normal information from image intensity, contour shape and the like.

These mathematical solids are a subset of the more general family of Generalised Cylinders[106].

Barr, A.H., *Superquadrics and angle-preserving transformations*, IEEE Computer Graphics and Applications **1** (1981), 1–20.

H.W. Hughes

269 Surface Reconstruction

Recovering object surface information can be achieved in a number of ways. Stereopsis[261] will yield explicit surface information only at those points where a direct correspondence is achievable. Analysis of motion or Optical Flow[190] also yields surface information but again the data is sparse. By using a 'thin plate' and deforming it to fit the available constraints it is possible to obtain an *approximation* to the object surface. By varying the resolution to which the thin plate is deformed it is possible to form a hierarchy of surface descriptions. At the lowest level these include texture descriptions while at higher levels coarser shape information is represented.

Terzopoulos, D., *Multilevel computational processes for visual surface reconstruction*, Computer Vision, Graphics and Image Processing **24** (1983), 52–96.

Terzopoulos, D., *Multiresolution Computation of Visible-surface Representations*, MIT Press, Cambridge, MA, 1984.

H.W. Hughes

270 Surface Segmentation

A form of Region Finding[230] that organises data from a range image into patches with a coherent property, usually based on some form of local surface shape. The two principal curvatures principal curvature or the mean and Gaussian curvatures are often used for purely data-driven segmentation; that is, all adjacent image pixels with similar curvatures are grouped to form a patch. Refinements include using only the signs of the curvatures to produce larger patches and to overcome noise, and prohibiting merging across depth discontinuities. Another refinement fits quadratic (or higher order) surfaces to stable patches and then incrementally grows the regions by adding image pixels that lie close to the estimated surface.

Besl, P., *Surfaces in Range Image Understanding*, Springer-Verlag, New York, 1988.

Robert B. Fisher

271 Symbolic Marker-passing

(*Non-numeric Spreading Activation, Von Neumann Machine Connectionism*)

Based on early work by Quillian (1966), symbolic marker-passing is used an efficient search which finds paths through a Semantic Net[244]. This search is usually characterised by being Breadth-first[28], massively parallelisable, and non-deductive. The canonical example of the need for marker-passing is found in the story 'John wanted to commit suicide. He picked up a rope.' To find that the context of the story is 'hang' requires the spread of information through the network, and the finding of the path: SUICIDE, KILL, HANG, NOOSE, ROPE. This formulation of marker-passing has been applied in natural language processing and planning research. Recent research has been examining integrating this technique with that of local connectionism. See also Marker-passing[156].

Charniak, E., *Passing markers: a theory of contextual influence in language comprehension*, Cognitive Science **7** (1983), 171–190.

James A. Hendler

272 Table of Multiple Effects

The Table of Multiple Effects (TOME) is a table (first used in the NOAH planner) relating patterns (representing facts) to nodes in a plan where they are asserted or denied. It is used by Non-linear Planners[181] to detect quickly interferences between actions on parallel branches. However, it can also be used to recognise beneficial side effects which may allow other goals to be satisfied without introducing new actions.

Sacerdoti, E.D., *A Structure for Plans and Behaviour*, Elsevier North-Holland, New York, 1977.

Austin Tate

273 Template Matching

A simple technique that is sometimes used in language understanding. A language unit (sentence or phrase) is compared with a set of predefined 'templates'. Some positions in the template consist of variables that match any input. If a template matches the input, the variables take the value of the corresponding elements in the input. For example the template 'Var1 hits Var2' matches the input 'Mary hits John with her hand' and Var1 takes the value 'Mary' and Var2

'John with her hand'. Template matching is only useful if there is a small number of templates, otherwise the matching process is too expensive. Template matching has been successfully applied to natural language generation too.

Barr, A. and Feigenbaum, E.A. (eds.), *The Handbook of Artificial Intelligence*, vol. 1, William Kaufmann, Los Altos, CA, 1981.

Maarten van Someren

274 Temporal Difference Methods

The temporal credit assignment problem is the problem of assigning credit to particular actions or events in determining the outcome of a process in terms of some measure of success. If for example I win a game of chess, how do I determine which moves in the course of the game contributed to my win? Normally we want to reward good moves (make them more likely) and punish bad ones (make them less likely).

Temporal difference methods are a class of learning algorithms which attempt to solve the temporal credit assignment problem. Temporal difference methods work by backpropagating reinforcements (measures of success) generated at any point in a process over chains of state-action pairs. By doing this the learner adjusts estimates of how successful particular actions chosen in particular states will be in the long term. Q-learning, and the AHC and bucket brigade algorithms are the most recent additions to the field of temporal difference learning. Temporal difference methods are also related to Dynamic Programming[82].

Sutton, R.S. (ed.), *Special Issue on Reinforcement Learning*, Machine Learning **8** (1992).

Jeremy Wyatt

275 Temporal Logic

Temporal logic deals with reasoning about time. There are three different approaches to temporal logic. The first approach is the Situation Calculus[253]. In the situation calculus, one simply adds an argument to each predicate which represents the time at which the predicate is assumed to be true. Thus, a two-place predicate like 'hit' becomes a three-place predicate, and the proposition that 'Harry loves Mary at time t' is represented simply as `hit(harry,mary,t)`. The second approach is the reified approach. In the reified approach, of which the Event Calculus is an example, one complicates the language by introducing separate terms for events, processes etc. as well as a predicate which is sometimes called `HOLDS`, sometimes TT, for saying that an event took place at some time. So, to express the proposition that 'Harry hit Mary at time t,' one would write `HOLDS(hit(harry,mary),t)`, where `hit(harry,mary)` is a functional expression denoting the event of 'Harry hitting Mary.' The final and third approach, which has not been too popular in AI, is modal temporal logic. One introduces a number of modal operators (see Modal Logic[165]), such as `P` (for past) and `F` (for future) and uses these to represent time-dependent information. For example, the information that 'Harry hit Mary in the past' is represented as P(hit(harry,mary)).

Reasoning about time is important in a number of areas in AI. In planning, for example, it is necessary to reason about the effects that an action will have on the world, and this involves reasoning about future states of affairs. Another example is natural language processing where one is concerned with extracting temporal information from the tenses of sentences.

Galton, A. (ed.), *Temporal Logics and their Applications*, Academic Press, London, 1987.

Turner, R., *Logics for Artificial Intelligence*, Wiley, New York, 1984.

Han Reichgelt

276 Theorem Proving

(*Automatic Theorem Proving, Mechanical Theorem Proving, Computational Logic*)

The goal of theorem proving is to use mechanical methods for finding proofs of theorems to perform automated reasoning. The implementation language is usually First-order Logic[208], or some other sound and complete formalism. The problem can be stated as follows: given a set of hypotheses (axioms and/or theorems), a set of rules (called the rules of inference) which allow new facts to be derived from old, a given formula (the goal) should be derived in finitely many applications of the rules of inference.

Besides mathematical reasoning, which is the obvious application, many problems can be transformed into theorem proving problems: question-answering systems, program-synthesis, program verification, state transformation problems and so on.

Some theorem proving techniques are based on Herbrand's theorem (e.g., Resolution[234], Connection Calculus[47]) but some attempts have been made using natural deduction-based, and induction-based systems (see Recursion Analysis[226]). The reference provides a very good collection of many of the early attempts (up to 1970) in theorem proving.

Siekmann, J. and Wrightson, G. (eds.), *Automation of Reasoning*, Springer-Verlag, Berlin, 1983 (published in 2 volumes).

Fausto Giunchiglia

277 Time-delay Neural Networks

A limitation of feed-forward networks is their inability to model time-dependent data. A proposed solution to this problem is the addition of several, time-delayed connections between a pair of units. In this way time-dependencies may be modelled. Training may be performed using the usual back-propagation of error algorithm (see Multi-layer Perceptorns[173]). The number of free parameters is reduced by constraining a set of connections differing only in delay to have the same weight—these constraints also impart a translation invariance into the system.

Waibel, A., Hanazawa, T., Hinton, G., Shikano, K. and Lang, K., *Phoneme recognition using time-delay neural networks*, IEEE Transactions on Acoustics, Speech and Signal Processing **37** (1989), 328–339.
Hirai, A. and Waibel, A., *Phoneme-based word recognition by neural network: a step toward large vocabulary recognition*, Proceedings of IJCNN-90, the International Joint Conference on Neural Networks, vol. 3, Ann Arbor, MI: IEEE Neural Networks Council, 1990, pp. 671–676.

Steve Renals

278 Time Logic Based Planning

Propositions, and actions and their component sub-actions, may be related to the time intervals over which they respectively hold and occur. This enables a substantial part of the planning process to be undertaken by a general temporal reasoner utilising a suitable Temporal Logic[275]. The combinatorial cost, however, may be high.

Allen, J.F. and Koomen, J.A., *Planning using a temporal world model*, Proceedings of IJCAI-83, 1983, pp. 741–747, also in Readings in Planning (Allen, J., Hendler, J. and Tate, A., eds.) Morgan Kaufmann, San Mateo, CA, 1990, pp. 559–565.
Tsang, P.K., *Plan generation in a temporal frame*, Proceedings of ECAI-86, vol. 2, 1986, pp. 479–493.

Jim Doran

279 Top-down Parsing

(*Hypothesis-driven Parsing*)

In trying to parse a string with a grammar, if one starts with the grammar and tries to fit it to the string, this is top-down parsing. For instance with a Context-free Grammar[54], one starts with expansions for the initial symbol, and builds down from there trying to find an expansion which will get to the symbols in the string.

Winograd, T., *Language as a Cognitive Process, Volume 1: Syntax*, Addison-Wesley, Reading, MA, 1983.

Henry Thompson

280 Transfer

One of the two mainstream approaches to machine translation consisting of parsing the Source Language (SL) text into some abstract representation (usually, some kind of annotated derivation tree), from which a lexical and structural transfer (specific to the language pair involved) is made in order to obtain a similar abstract representation in the Target Language (TL). A TL generation component then maps that representation into the TL text. The level of transfer may vary between systems, and can be anything from a syntactic tree with deep structure information, to a semantically annotated tree.

Compare with Interlingua[132].

Tucker, A.B., *Current strategies in machine translation research and development*, Machine Translation: Theoretical and Methodological Issues (Nirenburg, S., ed.), Cambridge University Press, Cambridge, 1987.

John Beavan

281 Tree-Adjoining Grammars

(*TAG*)

Tree-Adjoining Grammar (TAG) is a grammatical (and mathematical) formalism which constitutes a tree generating system (as opposed to string generating systems such as Context-free Grammars[54]). TAG postulates the existence of a finite set of *elementary trees* out of which bigger syntactic trees can be built. Elementary trees are divided into *initial trees* and *auxiliary trees*—initial trees represent minimal syntactic structures while auxiliary trees correspond to minimal recursive structures. Bigger trees can be built by means of the operation of *adjoining*. Adjoining takes a recursive structure (auxiliary tree) and 'inserts' it into another tree (either an initial tree or a tree derived by previous adjoining). Two important properties of TAGs which enable them to characterise the strong generative capacity of grammars (that is, their capacity to characterise the structural descriptions associated with sentences) are:

- Extended Domain of Locality: TAGs have a larger domain of locality than Context-free Grammars[54] and linguistic formalisms based on them (such as Lexical Functional Grammar and Head-driven Phrase Structure Grammar[114]). Thus, the dependencies between a verb and all of its arguments can be stated (locally) in one construction.
- Factoring Recursion from the Domain of Dependencies: The elementary structures are the domains over which (linguistic) dependencies such as agreement, subcategorisation, and filler gap, for example, are stated. Recursion is factored out from the domain of the dependencies but is reintroduced by the operation of adjoining which embeds recursive structures (auxiliary trees) into other trees. Thus, the long distance behaviour of some dependencies is accounted for.

TAGs belong to the so-called mildly Context-sensitive Grammars[55]. Thus, TAGs are more adequate for characterising various phenomena which require more formal power than Context-free Grammars[54] and formalisms based on them.

Joshi, A.K. and Schabes, Y., *Tree-adjoining grammars and lexicalized grammars*, Definability and Recognizability of Sets of Trees (Nivat, M. and Podelski, A., eds.), Elsevier, 1992.

Joshi, A.K., Vijay-Shanker, K. and Weir D., *The convergence of mildly context-sensitive grammar formalisms*, Foundational Issues in Natural Language Processing (Sells, P., Shieber, S.M. and Wasow, T., eds.), MIT Press, Cambridge, MA, 1991, pp. 31–81 (Chapter 2).

Special Issue on Tree-Adjoining Grammars, Computational Intelligence **10:1** (1994).

Nicolas Nicolov

282 Trinocular Stereo

A method for calculating a (generally sparse) depth map for a scene by corresponding (features of) three intensity images taken from different viewpoints.

The major advantages of using three images rather than two (as used in biologically motivated binocular stereo) are the rejection of many of the false correspondences which may arise in the two image case and the improvement in the accuracy of the estimated depth as a consequence of using the extra data.

Ayache, N. and Lustman, F., *Fast and reliable passive trinocular stereo-vision*, First International Conference on Computer Vision (1987), 422–427.

R.M. Cameron-Jones

283 Truth Maintenance System

(*Reason Maintenance System*)

A truth maintenance system (TMS) is used to record justifications for assertions. Such justifications can be used to generate explanations and to track down the assumptions underlying assertions. For example, in RUP every justification is a disjunctive clause of sentential (propositional) atoms and any such clause can be treated as a justification. RUP's TMS takes a set of such propositional clauses and performs propositional Constraint Propagation[52] to ensure that every assertion with a valid justification is in fact believed by the system (thus ensuring a deduction invariant). RUP's TMS also ensures that there is an entry on a contradiction queue for every propositional clause all of whose atoms are false.

deKleer, Johan, *An assumption-based TMS*, Artificial Intelligence **28** (1986), 127–162.
Jon Doyle, *A truth maintenance system*, Artificial Intelligence **12** (1979), 231–272.

D. McAllester

284 Typed Languages

In a typed language, every value is partitioned into one or more sets called types. Every legal expression denotes a value, and hence has a type. Some languages require typing information to be supplied explicitly, and the types of subexpressions in an expression are then checked for consistency by a type checker. Other languages make use of a type inference system to determine this information. Programming languages in which the type of every expression can be determined by static program analysis are said to be statically scoped. Languages in which all expressions are type consistent are called strongly typed languages. Strong typing guarantees that no program can incur a type error at run time, a common source of bugs. Standard ML[258] is an example of a strongly typed language.

Cardelli, L. and Wegner, P., *On understanding types, data abstraction, and polymorphism*, Computing Surveys **17** (1985), no. 4, 471–522.

Kevin Mitchell

285 Typed Preconditions

In early problem solving systems (e.g., STRIPS) and planning languages (e.g., PLANNER) the operators were given a single set of preconditions which were always interpreted as 'test if true, or subgoal to make them true'. The need to distinguish between the case in which the planner should only check if something was already true or should be allowed to add further actions into a plan to make the condition true was recognised as an important search control mechanism in POPLER. This led to two different types of precondition. Additions were also made to various pattern directed invocation systems which clustered alternative methods together and made a choice from them on the basis of some pre-occurring fact.

The utility of typed preconditions and their different properties for both hierarchic domain description and planner Search Space[260] control was investigated in the NONLIN planner which recognised four precondition types: SUPERVISED, UNSUPERVISED, USEWHEN and ACHIEVE.

Davies, D.J.M., *POPLER—implementation of a POP-2-based PLANNER*, Implementations of PROLOG (Campbell, J.A., ed.), Ellis Horwood, Chichester, 1984, pp. 28–49.

Tate, A., *Generating project networks*, Proceedings of IJCAI-77, vol. 2, 1977, pp. 888–893, also in Readings in Planning (Allen, J., Hendler, J. and Tate, A., eds.) Morgan Kaufmann, San Mateo, CA, 1990, pp. 291–296.

Austin Tate

286 Unification

A process, used in Resolution[234], for determining whether two expressions of the Predicate Calculus[208] will match. Terms of an expression in predicate calculus may be of one of three forms: constant, variable or function, where the last of these is made up of a function symbol applied to a number of terms. A substitution is a set of ordered pairs where the first element of each pair is a term and the second a variable. Applying such a substitution to a formula means that each variable appearing in the set of ordered pairs is replaced by the term which it is matched with in the substitution. Two expressions are unifiable if there is a substitution (the unifier) which when applied to both expressions makes them identical. A unification algorithm determines whether the given expressions are unifiable and, if so, finds a unifying substitution. It is usual in resolution based systems to specify that the substitution which is applied to unify two expressions is the most general such substitution, known as the 'most general unifier'. Thus the expressions lose as little generality as is necessary to make the resolution go through.

Nilsson, N.J., *Principles of Artificial Intelligence*, Tioga Pub. Co., Palo Alto, CA, 1980, and Springer-Verlag, Berlin, 1982.

Dave Plummer

287 Unification Grammars

Unification Grammar (UG) is the collective name for a group of linguistic formalisms and theories based on the idea of linguistic objects as partial information structures. Most UGs assume that linguistic objects are graph structures—sets of pairs of feature names and feature values, with the values possibly complex (see entry for Feature Structures[91]). The primary operation for combining such structures is Graph Unification[113]. However, grammar formalisms based on term unification, such as Definite Clause Grammars[65], are closely allied, even though term and graph unification were introduced independently (by Robinson and Kay respectively). As well as providing a sound understanding of how features in the sense of traditional grammar should behave, UGs appear to offer great flexibility in their computational properties, due to the monotonic (order-independent) nature of the operation of unification. Current research issues in UGs include the definition of more expressive feature logics for describing linguistic objects, and efficient algorithms for computing with these.

Gazdar, G., Klein, E., Pullum, G. and Sag, I., *Generalized Phrase Structure Grammar*, Basil Blackwell, Oxford, 1985.

Shieber, S.M., *An Introduction to Unification-Based Approaches to Grammar*, CSLI Lecture Notes no. 4, University of Chicago Press, Chicago, IL, 1986.

Johnson, M., *Attribute-Value Logic and the Theory of Grammar*, CSLI Lecture Notes no. 16, University of Chicago Press, Chicago, IL, 1988, pp. 103–155.

Pete Whitelock

288 Universal Graph Representation

A method of storing and organising a database of graphs that reduces some expensive (exponential) operations to less expensive (linear) set operations. The universal graph is a graph that is a single 'super' graph that contains all the graphs in the database as subgraphs. These subgraphs can be referenced as subsets of nodes in the universal graph. It is applicable to many applications requiring graph representation such as a Semantic Networks[244], chemical structures, networks, etc.

Levinson, R.A., *Self-organising retrieval system for graphs*, Proceedings of the AAAI, 1984, pp. 203–206.

Mohan Ahuja

289 Unsupervised Learning

In unsupervised learning the learner is given a set of input patterns, but does not receive a corresponding set of output patterns (as happens in Supervised Learning[250]). It therefore does not have any explicit representation of desired input-output pairs. The aim of unsupervised learning is to find regularities in the input set. This often simply done by grouping together similar input patterns. Different learning procedures use different measures of similarity. They also produce different representations of the regularities found in the input set. These include Discrimination Nets[76] and Feature Maps[240].

Unsupervised learning procedures are often based on statistical techniques. Conceptual Clustering[44], for example, is related to cluster analysis and self-organising feature maps are related to principal components analysis. One principal difference between unsupervised learning and statistical analysis is that the latter relies on all the data being present at the same time, i.e., at the start of learning. Because many unsupervised learning techniques learn incrementally they can converge to different partitions of the input space given different sequences of input patterns.

Significant work on unsupervised learning has been done by Kohonen, Grossberg, and Rumelhart and Zipser, among others.

Grossberg, S., *Neural Networks and Natural Intelligence*, MIT Press, Cambridge, MA, 1988.

Rumelhart, D.E. and Zipser, D., *Parallel Distributed Processing: Explorations in the Microstructure of Cognition*, vol. 1 (Rumelhart, D.E., McClelland, J.L. and The PDP Research Group, eds.), MIT Press, Cambridge, MA, 1986, pp. 151–193.

Kohonen, T., *Self-organisation and Associative Memory (Springer Series in Information Sciences, vol. 8)*, Springer-Verlag, Berlin, 1989 (3rd edition).

Jeremy Wyatt

290 Variable-valued Logic

Variable-valued logic is an extension of some known many-valued logics (MVL) in two directions:

- It permits the propositions and variables in the propositions to take values from different domains, which can vary in the kind and number of elements and also in the structure relating to the elements.
- It generalises some of the traditionally used operators and adds new operators which are 'most orthogonal' to the former.

Variable-valued logics have found applications in pattern recognition, medical decision making, and discrimination of structural textures. These logics have been successfully used in construction of diagnostic expert systems that can acquire knowledge by inductive learning from examples.

Michalski, R.S., *Variable-valued logic and its application to pattern recognition and machine learning*, Computer Science and Multiple-Valued Logic: Theory and Applications (Rine, D.C., ed.), North-Holland, Amsterdam, 1977.

Michalski, R.S. and Chilansky, R.L., *Knowledge acquisition by encoding expert rules versus computer induction from examples: a case study involving soya-bean pathology*, Fuzzy Reasoning and its Applications (Mamdani, E.H. and Gaines, B.R., eds.), Academic Press, London, 1981.

L.J. Kohout

291 Viewer-centred Co-ordinates

Objects described in viewer-centred co-ordinates, as opposed to Object-centred Co-ordinates[184], are those described in terms of a co-ordinate frame centred on the viewer, i.e., the viewer is considered to form the origin of the system and axes are defined relative to the line of sight and the retina. The representations of such objects must be transformed whenever the viewer moves.

Marr, D., *Vision: a Computational Investigation into the Human Representation and Processing of Visual Information*, W.H. Freeman, San Francisco, CA, 1982.

T.P. Pridmore, S.R. Pollard, S.T. Stenton

292 Viewpoint Determination

(*Model Parameter Determination*)

An important problem in model-based computer vision systems is to solve for the orientation and position of a three-dimensional object given some matches between features of the object and edges in the image. This algorithm provides a practical solution to the problem, including the difficult case of determining full three-dimensional viewpoint from locations of features in a two-dimensional image. The method is also capable of solving for internal model parameters such as variable sizes and articulations. The solution is based upon providing a rough initial estimate for the unknown parameters and then using Newton iteration to achieve an optimal least-squares fit. An important application is as part of a search process, in which a few matches can be used to determine viewpoint, which in turn constrains the locations of further matches.

Lowe, D.G., *Perceptual Organisation and Visual Recognition*, Kluwer Academic Publishers, Boston, MA, 1985.

David Lowe

293 Viewsphere Representation

A method of object representation in which the appearance (under orthographic projection) of the object is modelled over the sphere of possible viewing directions. As the object's appearance in intensity terms depends on the illumination, the form of appearance usually modelled is more stable, e.g., the visibility of a region or the topology of a line drawing of the object. The representation may then consist of the viewsphere regions in which the appearance remains constant and the corresponding appearances in those regions. Considerable interest is also shown in the boundaries between viewsphere regions, which often represent singularities of appearance.

Gigus, Z., Canny, J. and Seidel, R., *Efficiently computing and representing aspect graphs of polyhedral objects*, Pattern Analysis and Machine Intelligence **13** (1991), 542–551.

R.M. Cameron-Jones

294 Vocoder Representation

The vocoder representation characterises the speech signal in terms of energy in a series of fixed frequency bands. This form of analysis has been used as a stage in speech recognition, and as a means of data compression in storing speech for later resynthesis.

Linggard, R., *Electronic Synthesis of Speech*, Cambridge University Press, Cambridge, 1985.

Steve Isard

295 Vowel Quadrilateral

(*Vowel Analysis*)

The vowel quadrilateral technique employs an 8-pole Linear Predictive Coding[149] analysis to provide 9 vocal tract log area parameters which are mapped by a principal component transformation into a plane. The principal component coefficients are calculated off-line. It can provide a real-time display of vowel position, and a further rotational transformation gives a close approximation to the classical vowel quadrilateral.

Bristow, G.J. and Fallside, F., *Computer display of vowel quadrilateral with real-time representation of articulation*, Electronics Letters **14** (1978), 107–109.

Geoff Bristow

296 Willshaw Networks

The Willshaw network was designed to model the regular structure of nerve nets in the central nervous system. It is composed of a matrix of binary synapses (i.e., weighted connections) that feed into binary output units—whose function it is to threshold the sum of the input signals. The Willshaw network is used as an Associative Memory[12], i.e., it maps inputs onto outputs, and, as such, operates more efficiently than the Hopfield network[123].

The Willshaw net employs a training rule which is a variant of Hebbian learning. For each pair of association patterns, the appropriate bit patterns are simultaneously presented along input and output lines and synapses are turned on where input and output lines are coactive. (Synapses which are not used in pattern learning stay off.) During recall, the input pattern is presented as before, and each output unit sums the contribution from all activated synapses connecting it to active lines in the input pattern. In a fully connected Willshaw net, this sum is then thresholded according to the number of bits turned on in the original pattern, however, for partially connected nets, the thresholding strategy becomes more complicated.

Willshaw, D.J., Buneman, O.P. and Longuet-Higgins, H.C., *Neurocomputing: Foundations of Research* (Anderson, J.A. and Rosenfeld, E., eds.), MIT Press, Cambridge, MA, 1988.

Beale, R. and Jackson, T., *Neural Computing: An Introduction*, Adam Hilger, Bristol, 1990.

Ashley Walker

List of Contributors

Mohan Ahuja
Igor Aleksander
Bob Beattie
John Beavan
Martin Bennet
Hans Berliner
Alan W. Black
Robin Boswell
Max Bramer
Geoff Bristow
Paul Brna
Alan Bundy
R.M. Cameron-Jones
John Carroll
Alison Cawsey
Patric Chan
W.F. Clocksin
Don Cohen
Robert Corlett
Dave Corne
John Darlington
Rina Dechter
Roberto Desimone
Jim Doran
Mark Drummond
Janet Efstathiou
Roy Featherstone
Robert B. Fisher
Andrew Fitzgibbon
Ian Frank
Fausto Giunchiglia
P.M.D. Gray
Aloysius Hacker
Frank van Harmelen
James A. Hendler
H.W. Hughes
Steve Isard
Luis Jenkins
Karen Sparck Jones
Alistair Knott
L.J. Kohout
Robert Kowalski
Pat Langley
Andrej Ljolje
Jim Longstaff
Ashley Lotto
David Lowe
Helen Lowe
John Lumley
Seán Matthews
Jon Mayhew
D. McAllester
Chris Mellish
Martin Merry
Kevin Mitchell
Steve Muggleton
Nicolas Nicolov
M.J. Orr
Judea Pearl
Colin Phillips
Dave Plummer
S.R. Pollard
Kevin Poulter
T.P. Pridmore
Han Reichgelt
Alexander Reinefeld
Steve Renals
Dave Reynolds

Graeme Ritchie
Peter Ross
Geoffrey Sampson
Fritz Seytter
Alen Shapiro
Flávio Corrêa da Silva
Bernard Silver
Aaron Sloman
Alan Smaill
Maarten van Someren
S.T. Stenton
K. Sundermeyer
Austin Tate
Henry Thompson
Andrew Varga
Richard Waldinger
Ashley Walker
Mark Wallace
Lincoln Wallen
Toby Walsh
Martin Westhead
Pete Whitelock
Jeremy Wyatt
Richard Young

Index

Index of keywords via technique numbers.
Underlined numbers indicate a complete entry describing the keyword.

Springer and the environment

At Springer we firmly believe that an international science publisher has a special obligation to the environment, and our corporate policies consistently reflect this conviction.
We also expect our business partners – paper mills, printers, packaging manufacturers, etc. – to commit themselves to using materials and production processes that do not harm the environment. The paper in this book is made from low- or no-chlorine pulp and is acid free, in conformance with international standards for paper permanency.

Druck: Strauss Offsetdruck, Mörlenbach
Verarbeitung: Schäffer, Grünstadt